The Prospect of Immortality in Bilingual American English and Traditional Chinese
永生的期盼
美式英文—繁體中文雙語版本

I0094735

Front Cover © Sinclair T. Wang (王振祥)

i

The Prospect of Immortality in Bilingual American English and Traditional Chinese
永生的期盼
美式英文—繁體中文雙語版本

By Robert C.W. Ettinger
羅伯 艾丁格 著

WITH A NEW (2010) FOREWORD BY THE AUTHOR
內有新原作者前言(2010 年)

Sinclair T. Wang, Translator
王振祥，翻譯者

(Charles Tandy, Editor)
(唐萬龍，編輯者)

Ria University Press

www.ria.edu/rup

2010

Printed in the United States of America

Ria University Press   **Palo Alto, California**

iii

The Prospect of Immortality in Bilingual American English and Traditional Chinese
永生的期盼
美式英文—繁體中文雙語版本

By Robert C.W. Ettinger
羅伯 艾丁格 著

WITH A NEW (2010) FOREWORD BY THE AUTHOR
內有新原作者前言(2010 年)

Sinclair T. Wang, Translator
王振祥，翻譯者

(Charles Tandy, Editor)
(唐萬龍，編輯者)

FIRST PUBLISHED IN HARDBACK AND PAPERBACK 2010

PUBLISHED BY
Ria University Press
PO Box 20170 at Stanford
Palo Alto, California 94309 USA

www.ria.edu/rup

Distributed by Ingram
Available from most bookstores and all Espresso Book Machines

iv

The Prospect of Immortality in Bilingual American English and Traditional Chinese
永生的期盼
美式英文—繁體中文雙語版本

By Robert C.W. Ettinger
羅伯 艾丁格 著

In the 1960s Ettinger founded the cryonics (cryonic hibernation) movement and authored *The Prospect of Immortality*. Ettinger sees "discontinuity in history, with mortality and humanity on one side -- on the other immortality and transhumanity." Cryonic hibernation (experimental long-term suspended animation) of humans may provide a "door into summer" unlike any season previously known. Ettinger argues for his belief in "the possibility of limitless life for our generation."

在 1960 年代艾丁格創始了冰凍人(冰凍休眠)運動，並且著作了 *永生的期盼* 一書 。 艾丁格洞見了 "歷史上的不連續，一邊是會死和凡人—另一邊則是不死和超人。" 人類的冰凍休眠（實驗性的長期活體休眠）可能可以提供一道 "進入盛夏之門"，而且是一個不像任何前所熟悉之季節。愛丁格爲他在 "我們世代中無窮盡生命可能性" 的信念進行了辯解。

PREFACE BY THE TRANSLATOR

譯者前言

PREFACE BY THE TRANSLATOR
Dr. Sinclair T. Wang
BIONIN NANOBIOTECH, INC., CEO

While translating this book, I experienced the death of my father, and four months later that of my mother as well. The endless drives back and forth from my parents' home and the ceaseless visits to the hospital made real for me the helplessness and despair with which we must all face death as it is today.

In spite of what medical science has done to extend the human lifespan, the prospect of a certain end to life inevitably looms larger and larger as we age. At the end of the day, every human being still has to confront the reality of death. Its utter irreversibility and the unavoidable prospect of permanently severing emotional ties with loved ones is something that results in a terrible sense of helplessness and fear for everyone. It is, by any account, inhumane and cruel. But to make death a reversible choice would be to relieve this unnecessary and painful burden that we must all now bear.

譯者前言

王振祥 博士

拜寧奈米生技股份有限公司總經理

　　在翻譯這本書的過程中，我經歷了家父的逝世，四個月後，家母也相隨而去。其間無止盡地奔波於父母的家，以及無休止地進出醫院，讓我真實地感受到當今為止，我們都仍然必需面對的死亡中所含的無助與無奈。

　　雖然現代醫學進步，能夠延長人類的壽命，但是年紀逐漸邁老，死亡的陰影就不可避免地逐漸籠罩，終究還是得面對某一生命的終點。在日子的終了，每一個人還是得面對死亡的現實。每一個人對死亡之所以會有極度的無奈和恐懼感，乃因其是全然不可逆的，而且一定要和所有世上的感情牽繫永遠地切割。這不管如何總是一件多麼沒有人性而且是殘忍的事。如要消彌這種每一個人現在都必須忍受的莫須有而且是痛苦的重擔，唯有讓死亡變成是一種可逆的選擇。

To comfort my parents in their weakest, most helpless moments, I concluded some twenty years of residence in Seattle, Washington, and returned home to Taiwan. There, by the seaside, I would sometimes stroll through the groves of bamboo where I could take time observing the communal life of herons.

The bamboo canopy is always trembling with the noise of the birds busying themselves raising the next generation. But underneath, on the ground littered with decaying bamboo leaves, I could sometimes hear the skittering of dying herons, uttering feeble cries of pain. Then and there, I realized, Nature had destined them to die, making way for others and forcing a terrible contrast to the vigorous rebirth seen in the trees above.

But Nature may also give us inspiration. Technology has a great power to help humankind, but there must be a balance in its application. This balance will shift with the advances of the future, and has to be determined by careful, reasoned thought. While technology may be able to sustain life for a very long time, the quality of the life we live may dictate the necessity of coming to leave it behind at some point, like the herons do. For a mechanically and chemically sustained life can become not only meaningless, but also wasteful of resources for the succession of new life. Unless there is adequate wisdom to balance our emotion and reason, it will be difficult to make that decision properly.

在父母人生最孱弱無助的時光，為了要能夠慰藉他們，我結束了在華盛頓州西雅圖居住二十幾年的生活，回到了我的家鄉台灣。在那裡，我偶而會抽空獨自躑躅在濱海的竹林中，觀賞鷺鷥的群居生活。

竹林頂層往往是充滿鷺鷥們忙於養育下一代的陣陣喧囂聲。然而在竹林底下，散滿枯萎腐爛竹葉的土地上，我時時可以聽到瀕死掙扎的鷺鷥所發出痛苦的虛弱哀嚎。就在那當下，我深知大自然就要命定它們死去，騰出空間給其它鳥。和林梢所見活躍的新生，這真的強逼出一個何等慘烈的對比。

從自然界我們或許可以得到一些靈感。科技具有強大的能力來幫助人類，但是在其實際應用上應該有一個平衡點。此平衡點會隨著未來的進步而有所移動，必須要靠小心理性的判斷來拿捏。雖然科技能夠維持生命許久，但是我們存活的品質會決定到達某一點時，或許我們就應該像鷺鷥般地放棄生命，因為靠機器和化學物質維持的生命會變成不僅沒有意義，也會虛耗繼起新一代生命的資源。除非有足夠的智慧來平衡我們的感性和理性，否則要下一個恰到好處的抉擇是非常困難的。

Freezing technology and prospect of future revival and rejuvenation may make such an action easier. When death becomes a reversible choice, making that decision will not entail so much uncertainty. By then, even suicide will go through a morphological transformation and ethical reevaluation. When one becomes too sick or tired of life, there'll be no need to choose permanent death. One can adopt suspended animation to enjoy a temporary escape, and choose to return to life when this reality has been transcended.

Throughout the history of human civilization, humans have used technology to solve the problems that hold us back, especially the limitations of time and distance: modern drugs have significantly extended the human lifespan; the jet plane makes travel between continents an everyday phenomenon; and computers and the internet have eliminated physical separation as barrier to communication between individuals. Now, coming advancements in biotechnology, nanotechnology and artificial intelligence, such as human cloning, organ and tissue engineering, gene therapy, brain-computer interfaces, and more, all indicate that the indefinite extension of life is near.

有了冰凍技術和未來復甦回春的盼望，下定這種抉擇可能會容易許多。當死亡變成是一種可逆的選擇時，下定這種抉擇就不再會牽扯到那麼多的不確定性了。到那時，就連自殺都會有一個形式上的轉變和倫理道德上的重估。當某個人已經病到藥石無效或是對生命已經感到厭倦無趣時，他不需去選擇永久性的死亡。他大可以去採行冰凍休眠來享受暫時的解脫，等到他的現實已經轉化後，才選擇來重回人世。

在整個人類文明的歷程中，人類一直都在利用科技來解決存在的無奈，尤其是對時間和空間的限制：例如現代醫藥明顯地延長了人類的壽命；噴射機讓洲際間的旅行變成是每天稀鬆平常的事情；而電腦和網際網路讓人與人間溝通上時空隔閡的障礙完全消失。當下，在未來生物科技，奈米科技和人工智慧的進展，例如人體複製，器官和組織複製工程，基因療法，人腦–電腦介面技術等等，在在都明顯地指出壽命無限延伸乃是近在眼前。

From history, we know that new concepts and technology are usually subject to a period of persecution by the conservative elements of society before they are accepted. For example, science had to struggle for almost two hundred years before it was accepted that the solar system is not geocentric, and now routine in vitro fertilization was condemned by many when it first appeared. Judging from past evidence, I believe that when comprehensive legal and ethical regulations are ready, embryonic cell research, human cloning and eventually immortality will gradually come to be accepted by all of human society, including those that believe in the divine creation.

While the emphasis of the author, Mr. Ettinger, is on evaluating the feasibility of human cryopreservation, his thoughts on legal, ethical, religious and socioeconomic issues also apply more generally to any world where immortality is realized, regardless of method. Though these issues are, in some sense, beyond pure scientific reasoning, their resolution is important before a majority consensus can be achieved, codes and regulations can be ratified, and the technology can finally be embraced by society. However, following history, they have no absolute and invariable answers. Instead, they will continue to evolve with the advancement of science and the gathering of new knowledge.

從歷史上我們得知，新觀念和科技在被接納之前往往都要遭受到社會中保守勢力一段時間的逼迫。例如科學界要虛耗約兩百年的時光才能讓地球不是太陽系的中心被完全接納；而當今被例行應用的試管授精技術，起初出現時也蒙受許多人的詛咒。從過往證據的判斷，我深在法律和倫理道德的配套規範齊備之後，信胚胎細胞的研究，人體的複製，以及最終人類的不死，也都會逐漸被整個人類社會所接受，甚至於包括那些相信神聖的創造論者。

Ettinger 先生著作這本書的重點雖然在論述冰凍人的技術可行性，然其在法律，倫理，宗教和社會經濟等議題上的著墨，其實也涵蓋了人類實現永生後的任何世界，不管其獲得永生的路徑為何。這些議題雖然在某些方面是超越純粹科學的理性和邏輯，但是在取得多數人的共識，通過法律條文和規範，以及此科技最終可以被社會接納之前，能夠得到其結論是非常重要的。然而依據歷史，它們並沒有絕對的和不變的答案。它們反而是會隨著科技的進步和新知識的累積而繼續地進化的。

The time it takes for the adoption of these new ideas will have a direct impact on human welfare; belated acceptance will reduce the number of people who can enjoy immortality. Immortality proponents should not, however, become hasty and attempt to force it on an intellectually and mentally unprepared society. This idea calls for more debates, communication, education and even patient preaching. The appearance of this book in Chinese is to further this goal with the huge number of the world's population who use that language.

When Mr. Ettinger wrote this book in the 1960s, the computer industry was just beginning to develop, genetic bioengineering was barely on the horizon, and no one had even heard of nanotechnology. But in his book, he had foreseen the ultimate endpoint of those technologies, and among his predictions there are many that have already become reality. One has to admire his foresight, and I believe it gives all the more credibility to his other, as yet unrealized predictions.

Perhaps due to the atmosphere of the Cold War era, Mr. Ettinger does seem to have let his position as an objective scientist be swayed ever so slightly by politics when it comes to ideas about Mao and Communist society. Out of respect to those readers living in a post-Mao society, I have to mention my regret at these negative comments.

只是接納這些新觀念所耗費的時間會直接影響到全人類的福祉；越延遲的接納將會讓能夠早日享受到永生的人數減少。然而永生推動者也不能因此而變得躁進，強把此加諸於一個心智尚未準備好的社會。這個觀念還是有賴於更多的辯論，溝通和教育，甚至於耐心地說教。這本書以中文的版本再出現也是想要進一步和全世界為數浩大使用此語言的人口達到前述功能的目標。

Ettinger 先生是在 1960 年代著作此書，當時電腦產業的開發剛開始萌芽，生物基因工程幾乎尚未出現在地平線，而奈米科技更是前所未聞。但是在其書中，他似乎都已經預先看到了這些科技發展極致後的景象，其中還有許多他的預言當今已經成為事實。在此我們不得不佩服其遠見，因此我相信這也會讓人對其在其他尚未成真的預測給予更高的信用度。

Ettinger 先生可能是身處在冷戰時代的氛圍，在攸關毛澤東和其共產社會上的觀念，他似乎有讓他身為一個客觀科學家的地位稍許受到政治的影響而搖移。為了尊重那些目前還生存在毛後社會的讀者，我不得不對這些負面的觀點稍微表達我的遺憾。

Politics is, of course, not strictly science. Behind it are intangible historical burdens and an ever-complicated thicket of entangled cause and effect. It is therefore hard to jump to an absolutely right or wrong conclusion. Each community has been given a different reality and set of limitations, and from these conditions naturally adopts a social system that best fits its place in the world. The contribution of science and technology is to help us understand each community's state as a result of varying conditions in transition, and to resolve potential conflicts that may arise from a difference in understanding, so that a commonwealth may soon be realized on earth.

Notwithstanding that, Mr. Ettinger is a thoroughly respectable scientist, and I greatly admire his work. It is an extraordinary honor and pleasure to have the opportunity to translate this book for a new generation of readers, and I hope that it will spark for them further discussion and debate about the ideas contained within.

政治本來就是非純粹的科學。而在其背後有不可數算的歷史包袱和永遠複雜叢生、糾葛不清的因果關係。因此絕對性的是非對錯難以遽下結論。每一個族群都被給予一套不同的現實和限制條件，而從這些條件自然地會選擇出一套其安身立命最合適的社會制度。科學和技術的貢獻乃是在由過渡時期中所衍生的各種變化條件中，去理解每一個族群所處的狀態，並且去化解可能有彼此理解之間的差異所可能產生的各種衝突，因而世界大同的理想方有可能早日實現。

　　不管如何，**Ettinger** 先生徹頭徹尾是一位可尊敬的科學家，我對他的著作極其推崇。能夠有機會為新一代的讀者翻譯此書是我無上的榮幸和快樂，而我希望這也能激發出他們對書中諸觀念進一步討論和辯論的火花。

FOREWORD 2010 — ROBERT ETTINGER

2010 年前言 — ROBERT ETTINGER

FOREWORD 2010 — ROBERT ETTINGER

Since the first commercial version of *The Prospect of Immortality* (Doubleday, 1964) no one has lived forever, and since the first edition of *Man into Superman* (St. Martin's Press, 1972) no one has leaped a tall building, but those works have held up pretty well in terms of relevant technology. In terms of conjectures about the future, *Superman* is still far ahead of its time, and *Prospect* was proven ludicrously over-optimistic about the rate of growth of the cryonics movement. Nevertheless, progress has been made and the wind is at our backs.

Member profiles: Cryonics Institute (CI) members tend to be better-educated than average, but come from many segments of society. There are more men than women. We have members/patients in/from Australia, Austria, Canada, England, France, Germany, Norway, Singapore, and Sweden, as well as the U.S. We work with members at a distance – in the U.S. and abroad – to set up local emergency arrangements.

Member and patient numbers: As of the end of September 2009, CI had over 800 members (not all funded) and 95 patients. (See web site <www.cryonics.org> for more detail.) Alcor had over 900 members and 88 patients. Most of Alcor's patients are "neuros" or head-only. CI's are all whole body, with two notations. One, a few neuros that had been previously processed and stored elsewhere, and later owing to closing of organizations had nowhere else to go, were accepted for storage at CI. Secondly, CI's current vitrification process is applied primarily to the head, with a different protocol for the rest of the body. But CI continues not to offer a "neuro" option, primarily because of bad public relations, especially intra-family.

2010 年前言 — ROBERT ETTINGER

自從*永生的期盼*一書首度商業版發行至今**(Doubleday** 出版社，**1964)**，沒有一個人變成永生不死，而且自從*人變成超人*一書初版的發行至今**(St. Martin** 出版社，**1972)**，也沒有一個人可以從高樓一躍而下，但是就其相關科技而言，這些努力還是持續不斷。從對未來的臆測來講，*超人*還是言之過早，而*期盼*則被證實爲對冰凍運動的成長率有點可笑地過度樂觀。然而不管如何，還是有所進展，而且目前我們已是順風而行。

會員剖析：冰凍機構 **(Cryonics Institute, CI)**的會員平均教育水準高於一般人，但是是來自於社會不同的區塊。男人數目比女人多。我們的會員或病人是在或來自澳洲，奧國，加拿大，英國，法國，德國，挪威，新加坡，瑞典，以及美國。在美國以及在國外，我們會和會員進行遠距離的合作，來架構起一些地區性緊急事宜的處置。

會員和病人數目：到 **2009** 年 **9** 月底，**CI** 有超過 **800** 個會員（不是全部都有繳費），並且有 **95** 個病人。**(**進一步細節請參閱網站**<www.cryonics.org>)**。**Alcor** 有超過 **900** 個會員，並且有 **88** 個病人。大部分 **Alcor** 的病人都是"神經體**(neuros)**"，也就是只有頭部。**CI** 除了有兩個特例之外，全部都是整個身體。一個是先前在別處處理和儲存的幾個神經體，之後因爲該機構的關閉無處可去，而被接收儲存在 **CI**。第二是 **CI** 目前的玻璃化程序主要是應用在頭部，身體其餘部分則依另一不同的處理程序。但是 **CI** 還是堅持不提供"神經體"的選項，主要乃是因爲其不良的公共關係，尤其是在家族內部之間。

Nasty book: In October of 2009 a book was published by Larry Johnson and Scott Baldyga called *Frozen: My Journey into the World of Cryonics, Deception, and Death*. The book got lots of media attention, Johnson having several years earlier been for a time the COO of Alcor Life Extension Foundation. Its effect on Alcor and cryonics as a whole remains to be seen, but at this writing I see no evidence of any serious injury to Alcor or to cryonics generally, despite considerable forebodings and hand-wringing.

The thrust of the book is that (a) cryonics is a sham and a scam, but (b) should nevertheless be regulated like any "industry" with the good separated from the bad, and (c) Alcor has committed various crimes and unethical practices, some of which Johnson claims to prove through clandestine audio recordings and personal observation. A lot of attention is given to the cryopreservation by Alcor of baseball great Ted Williams.

It seems clear enough that Johnson is a scoundrel who lied, stole, violated his trust, and by his own account failed repeatedly to take timely action against the practices he later assailed. It seems likely that Alcor did some pretty stupid things, although to the best of my knowledge nothing remotely criminal or ill intentioned, and certainly nothing involving financial profit to individual employees.

My latest book: In another attempt to explain the philosophical and other foundations of cryonics, in 2009 I published another version of *Youniverse* with a new subtitle, *Toward a Self Centered Philosophy of Immortalism and Cryonics*.

噁心的書籍：在 2009 年 10 月一本由 Larry Johnson 和 Scott Baldyga 所出版的書書名為冷凍：*我進入冰凍技術，欺騙和死亡世界的歷程*。此書媒體曾極為注目，Johnson 先前有很多年曾任 Alcor 生命延長基金會的營運長。其對 Alcor 和冰凍術的影響整體而言還是有待觀察，但是當在撰寫此文時，大體上我是沒有看到對 Alcor 或是對冰凍術有任何重大傷害的蛛絲馬跡，縱然曾經有相當程度的不祥之兆和不知所措。

此書的攪局點是 **(a)** 冰凍術是一種欺詐和騙局，但是 **(b)** 不管如何還是必需要如其他"產業"一樣優劣有分地管制，並且 **(c)** Alcor 曾經犯下許多罪行和不道德的手法，其中某些部份 Johnson 宣稱有秘密錄音和個人觀察的證據。很多注意力是集中在 Alcor 冰凍儲存棒球名將 Ted Williams 的事。

似乎很清楚地可以看出 Johnson 是一個卑鄙小人，他欺騙，偷盜，違背誠信，而且自己多次沒有對他之後在責罵的作業方式採取及時的糾正動作。有可能 Alcor 曾經做出一些頗為愚蠢的事情，但是就我所知範圍內，其實也沒有犯罪或是惡意可言，更沒有牽涉到金錢利潤入員工個人口袋之事。

我最近之著作：另外位了要解釋冰凍的哲學和其他基金會，在 2009 年我出版了另一版本的 *你的宇宙 (Youniverse)*，其中加了一個次標題，*朝向永生主義和冰凍術的自我中心哲學*。

First I assert, and claim to prove, that me-first and feel-good are the necessary and only possible bases for consciously motivated thoughts and behavior. Although a few philosophers, down from antiquity, have agreed with this, the majority have not and do not, and on their side are massive influences of evolution and conditioning or cultural inertia or cultural legacy.

One way to look at the issue is the "spiritual" need to find "meaning" in life, including some high principle as grounding. I try to show that the usual principles, such as a religion or ideology or cultural legacy, are wrong and often counterproductive to the benefit of the individual. What I suggest instead is personal *integrity*, the insistence on realism and facing the truth, including the truth of uncertainty. I concede that buying into delusion may make some people happier or less discontented than they might otherwise be, but some of us refuse to settle for blind comfort.

Another contribution of the new book is a new perspective on the question of survival criteria or personal identity. Again, this is not easily made succinct, but the main idea is a quantitative view of identity and a physical overlap of past, present, and future qualia, rigorously allowing at least some degree of survival over time or identity of past, present, and future selves. We do not *have* qualia, but rather we *are* our qualia.

Suspended Animation Inc. (SA) and the CI web site: Several years ago CI entered into a contract with SA and began promoting SA to CI members and prospective members. That, in my opinion, was a very bad mistake, which has now been largely corrected. I believe SA has nothing to offer which cannot be obtained otherwise at much lower cost, and that SA's future is emphatically in doubt.

首先我假設並且宣示可以證明，自我優先和感覺爽快是能夠有意識地激勵思想和行為的必要而且是唯一可能的基礎。縱然從古至今有一些哲學家認同此點，但是多數並非如此，而且他們這一方對進化和薰陶，文化慣性或是文化遺留是有巨大影響力的。

　　看待此議題的另一方法就是以去找尋生命"意義"的那種"心靈上"的需求來當作基礎，包括某些高尚的原則。我想要表達的是那些一般的原則，例如宗教或是思想體系或是文化遺留都是謬誤的，而且往往對個人的福祉是反作用的。我所建議來替代的就是人格的"*完整*"，對實存的堅持並且面對真實，也包括不確定性的真實。我承認接受迷幻比不接受可能會讓某些人更快樂或是減少不滿足，但是我們之中的某些人就是拒絕接受盲目的舒適的。

　　此本新書的另一項貢獻就是對存活的標準或是個人特性上的問題提出一個新的觀點。再度強調，這是不容易簡練地表明的，但是其主要觀念就是對人格特質的量化觀以及一個針對過去，現在和未來特質的實質疊疊，至少嚴謹地容許某一程度的跨時間，或是跨過去，現在和未來的自我特質的存活。我們不是*有*特質的，而是我們*是*我們的特質。

活體休眠公司 (Suspended Animation Inc., SA) 和 CI 網站: 幾年前 CI 和 SA 簽訂一個合約，並且開始推薦 SA 給 CI 的會員和可能會員。我個人認為這是一個極大的錯誤，這現在已經有被大力地矯正了。我相信 SA 所提供的沒有什東西不可以以較低的成本從其他地方獲得，而且 SA 的未來很明顯地是堪慮的。

Among other things, it depends on heavy subsidy by Saul Kent and William Faloon through the Life Extension Foundation (LEF), a purveyor of vitamins and nutritional supplements. Kent and Faloon are also involved in for-profit companies including Twenty First Century Medicine (21CM), Critical Care Research (CCR), and Timeship. I have not been able to obtain documentation about the legal arrangements in place regarding successors and policies upon the death or incapacity of Kent or/and Faloon. SA has a glaring lack of any credible business plan.

Accelerating progress: Even though growth of the cryonics organizations has been much slower than I previously hoped and expected, the rate of growth has improved. CI dates back around 33 years, but roughly half of our members and half of our patients have come in the last five years or so, and if I remember correctly this is also true of Alcor. Trans Time is dormant as far as I know, and it appears the American Cryonics Society (ACS) is not as active in expanding its membership as it might be. Check their web sites for possible updates. KryoRus is a recent entry in Russia, which seems to have several patients. CI, Alcor, and ACS are all long-established non-profit organizations legally recognized and incorporated as such.

除此之外，他是通過生命延長基金會 (Life Extension Foundation, LEF) 仰賴由 Saul Kent 和 William Faloon 提供的重大補助，這是一家維他命和營養輔助品包商。Kent 和 Faloon 也和一些營利公司有關，其中包括二十一世紀醫藥 (Twenty First Century Medicine, 21CM)，緊急照護研究 (Critical Care Research, CCR) 和時間船 (Timeship)。針對現有有關繼承人和在 Kent 和/或 Faloon 死亡時或是無行為能力時處置上的法律安排的文件，我還沒有辦法去取得。SA 很明顯地缺乏一個有任何可信度的商業營運計畫。

加速進展：雖然一些冰凍組織的成長比我先前所期望和預期的要慢上許多，但是現在其成長速度已經有所進步。雖然 CI 的成立可以追朔到大約 33 年前，但是大約我們一半的會員和我們一半的病人都是再最近五年左右參與的，而且假使我記得沒錯的話，Alcor 的情形也是如此。據我所知 Trans Time 是處於休業狀態，而美國冰凍協會 (American Cryonics Society, ACS) 在會員的擴增上則沒有如其本來應該般地活躍。可以查閱他們的網站來看有任何的更新。KryoRus 是俄羅斯的一個新成立機構，看起來似乎有許多病人。CI，Alcor 和 ACS 都是已經設立許久的非營利機構，法律上被認定而且也是依法來組成公司的。

The reason for accelerating progress seems clear enough and bodes well for the future. It is simply that, every year and almost every day, advances in science and technology tend to make our thesis more credible. There is plenty of objective evidence for this. As one example, when CI was founded it was almost unheard of for physicians and scientists to take seriously the prospect of an end of senescence, a cure for aging. Now there are several organizations and many eminent physicians and scientists boarding the anti-senescence bandwagon. As for the general public, the nutritional supplement business, offering sales of products claiming to extend life or prevent disease, has grown markedly even in the recent economic downturn. Barring catastrophe of one sort or another, it looks like onward and upward, excelsior!

None of this means we can afford to be complacent. We need growth and improvement all along the line. And every individual can make a contribution, starting with your self, family, and friends.

Robert Ettinger   (January 1, 2010)

會加速僅展的原因似乎非常清楚，而且未來也是看好的。這只是因為每年而且幾乎是每一天中科學和科技的進步，都傾向於讓我們的論點更加可信。針對此實在有太豐富的客觀證據。舉一例而言，**CI** 初成立時，對終結衰老和治癒老化的未來可能，幾乎未曾聽到有醫生或是科學家會認真以對。而現在有許多機構和多位傑出的醫生和科學家都搭上了抗衰老的車隊。 就一般大眾來看，營養補助品的生意，提供宣稱可以延長壽命或是預防疾病產品的販售，甚至在最近經濟惡化之下，還是有明顯的成長。撇開種種災難不談，一切看來好像有向前和向上在進展，棒極了！

這一切並不意味著我們可以來自滿。我們在整條路上都需要成長和進步。而每一個個人都可以有所貢獻，從你自己開始，到你的家族，然後到你的朋友。

Robert Ettinger   (**2010** 年 **1** 月 **1** 日)

CONTENTS

目錄

(Robert Ettinger)

CONTENTS

(Robert Ettinger)

目錄

xli

PREFACE

by Jean Rostand

de l'Academie francaise

前言

Jean Rostand

法國國家科學院院士

PREFACE

by Jean Rostand

de l'Academie francaise

About a century ago, Edmond About, a fine French writer and one of the precursors of "science fiction," published a short novel called *The Man with the Broken Ear*. In this diverting tale, he tells about a professor of biology who dries out a living man and then, after a "suspension of life" lasting several decades, successfully resuscitates him.

What was, in 1861, only an amusing fantasy has in our time taken on a rather prophetic air; for, in the light of recent scientific developments, a similar method of preserving a human being no longer seems so impossible.

We have learned, from the experiments of Hahn de Becquerel and others, that some animals of the lower orders (Rotifera, Tardigrada, Anguilla), some vegetable seeds and some microbes can have all internal activity interrupted for a long time by being reduced in temperature to close to absolute zero-and then, upon being thawed, resume all normal functions again. But more than this, researchers report having observed "resurrections" of this sort even among higher order animals; though the entire animal may not have been involved, it is definitely the case that a significant amount of tissue-and even whole organs-were thus frozen and revived.

前言

Jean Rostand

法國國家科學院院士

大約在一個世紀之前，法國有一個優秀的作家和"科幻小說"的先驅者 Edmond About，他出版了一本名為 *"耳朵壞掉的人"* 的短篇小說。在這個娛樂性的故事中，他描述了一個生物學教授，他曾將一個活人乾燥，然後經過了一段好幾十年 "生命暫停" 的時間之後，成功地將他復甦過來。

在 1861 年，這個故事聽起來雖然像是一個逗趣的天方夜譚，但是到我們這個世代時，卻帶有一種頗有先知的味道；因為如果以最近科學的一些進展看來，用一種類似的方法來保存一個人，似乎已經不再是那麼地不可能。

從 Hahn de Becquerel 和其他科學家的實驗中我們已經知道，某些較低綱目的動物 (輪蟲類，緩步動物類，鰻魚類)，某些植物的種子以及某些微生物，在接近絕對溫度零度的溫度下，其內在的生命活動都可以被暫時中斷一段很長的時間，然後在經過解凍之後，都可以再度恢復其正常的生命機能。況且除此之外，依據某些科學家的研究報告，他們觀察到類似這種 "復活" 的現象，甚至於也可以在較高綱目的動物身上發生。雖然這還不能涵蓋到整個動物界，但是可以確定地，大多數的組織－甚或是整個器官－可以被如此地冰凍後然後再使其復活的。

In the same way, the sperm of certain mammals, when impregnated with proper preservatives, has been able to endure the temperature of liquid nitrogen for some months without losing the ability to regain normal mobility and the capacity to reproduce. Likewise, the heart of a chicken, after undergoing a similar super cooling, was able to heat again after being rewarmed.

So it is not out of the question to anticipate future successes of greater and greater complexity; indeed, we are at last even forced to concede the real possibility that the means for freezing and resuscitating human beings will one day be perfected, at however distant a time this may be. This certainly is the opinion of M. Louis Hey, one of the most competent contemporary biologists in the field. He writes:

"There are some very convincing reasons to think that, thanks to future research, one will be able to bridge the gap that now separates the superior organisms from the Tardigrada and Rotifera; the solution will then be found to the problem of suspending the vital life force perhaps indefinitely." (Conservatism de 1a vie par le froid. Hermann, 1959.)

In *The Man with the Broken Ear*, Edmond About envisioned, with a certain amount of humor, some of the consequences for human society which could result from the preservation of human beings.

"The sick people who were declared incurable by the ignorant scientists of the nineteenth century need no longer bother their heads about it; they were dried up to wait peacefully in the bottom of a box until the doctors had found remedies for their ills."

一樣地，某些哺乳類動物的精子，如果注入恰當的防腐劑，就可以抵擋液態氮的溫度好幾個月，而不會失卻其恢復正常活性的能力以及繁殖複製的可能。同樣地，在經過相似的超低溫冰凍過程後的雞心臟，在其被再度地回溫後，也可以恢復其正常的跳動。

因此預期未來能夠在越來越複雜的生物體上成功，也絕對不是緣木求魚的事；的確地不管這還需要多久，到最後我們甚至會被逼著去接受冰凍和復甦人類的技術總有一天將會達到其完美境界的真實可能性。這當然也是當代在此領域最傑出能幹的生物學家 M. Louis Hey 的看法。他表示:

"有一些非常令人信服的理由來作如此的思考，藉著未來的研究，我們將可以來彌平現在介於高級生物和輪蟲類以及緩步動物類之間的鴻溝；如何可能無限期地暫停關鍵性生命力上的問題，藉此將可以因而得到解決。" (Conservatism de la vie par le froid. Hermann, 1959.)

在 "耳朵壞掉的人" 的小說中，Edmond About 以些許幽默的心態，預先看到人類的保存技術，所可能對人類社會帶來的一些因果。

"那些被一群十九世紀的蒙古大夫宣佈為回天乏術的病人，可以不用再費盡心思到處尋覓再世華陀了；他們都會先被乾燥起來，平心靜氣地在一個保存箱的箱底等待，直到這些醫師找到治癒他們的疾病的方法。"

R. C. W. Ettinger, the author of *The Prospect of Immortality*, has gone a crucial step beyond the French writer: It is not only the incurables he proposes to preserve, but the dead themselves. Indeed, as Mr. Ettinger suggests, should not the dead be considered to be only "temporary incurables" that a better informed science might one day resuscitate by repairing the ills to which they had succumbed -whether their difficulty be sickness, accident or old age? The preservation he advocates would be through refrigeration (a liquid helium or nitrogen bath); this is a method of freezing that is not harmless now, but undoubtedly the science of tomorrow will have ways of repairing freezing damage too.

So we don't have long to wait before we shall know how to freeze the human organism without injuring it. When that happens, we shall have to replace cemeteries by dormitories, so that each of us may have the chance for immortality that the present state of knowledge seems to promise. At the moment, all of this may seem like a remote chance, and no one is more aware of this than Mr. Ettinger. But he has the insight to realize that we have nothing to lose and, possibly, everything to gain by pressing the search. It is, in a sense, a Pascal's wager based on a faith in science. Certainly, a decision to let all corpses remain corpses is, in the face of Mr. Ettinger's alternative, the highest folly.

What is important to realize is that Mr. Ettinger is, in the strictly biological section of the book, carrying to its logical conclusion an argument for which he has unimpeachable premises.

*永生的期盼*一書的作者 R. C. W. Ettinger 教授，他超越了這個法國作家，向前邁進了一個關鍵性的步伐：他不僅倡議要將無藥可醫的人保存，而且是要將已乘鶴歸西的人本身保存。的確，就如同 Ettinger 教授所倡議的，當未來有一天更為先進的科技能夠將死去的人，藉著修復其所罹患的疾病來使其復活－不管其死因是疾病，意外事件或是老化，難道我們現在不應該將這些亡者認定為僅是"暫時不治"嗎？他所提倡的保存方法乃是透過冰凍 (一種液態氦或是液態氮的浸泡法)；這雖然不是一種在當下不具傷害力的冰凍法，但是無庸置疑地，未來的科技也一定有辦法來修復其冰凍傷害。

因此我們不需要等太久，就將知道如何沒有傷害地來冰凍人類這種生物。當一切成真時，我們就將會以一些停留所 (dormitories) 來替代所有的墳場，如此一來，我們每一個人才有機會來獲得在目前科技知識水準中所似乎在應許的永生。從目前看起來，所有的這些似乎都是遙不可及的，而且沒有人比 Ettinger 教授更加認清到這個事實。但是他卻能洞察機先，知道我們是無後顧之憂的，而且還有可能藉著對推動研究而左右逢源。從某個觀點看來，這似乎有點類似 Pascal 基於對科學的信心，所下的一個賭注。不可諱言地，在面對 Ettinger 教授的替代方案，如果還決定讓所有的屍體一直維持為屍體，真的會是天下最終極的愚癡。

很重要地，我們必須要認知到，Ettinger 教授在其書中於純粹屬生物科學的部分，為了讓其演譯出來的結論和論點合乎邏輯，他是有他的一些無懈可擊的假設前提的。

It is not the role of the prefacer to pronounce on the immediate practicality of the program. Indeed, Mr. Ettinger himself fully understands that the whole job cannot be done overnight. What he is telling us is that we must begin; the job will be done some day, and for every day that we put it off untold thousands are going to an unnecessary grave.

In any case, Mr. Ettinger's book is a captivating, stimulating tonic crammed with original views-especially on the problem of the personal identity of the individual. It deserves to be read and thought about.

Translated by Sandra Danenberg

然而去宣告這種計劃的立即可行性，當然不是我這個寫前言者該扮演的角色。**Ettinger** 教授自己的確也知道這整個計劃當然不可能是一蹴可及的。他所要傳達的訊息乃是我們一定要著手開始；這個工作在某一天一定會達成的，只是我們每多一天的延宕，就會有好幾千人要無謂地走入莫需有的墳墓。

　　不管如何，**Ettinger** 教授一書乃是充滿原創觀點，既扣人心弦又激發思緒的一本有營養的著作－尤其是在其有關個人的身分認定的問題上。這本書實在是值得你來讀來思考的。

　　原法文轉英文由 **Sandra Danenberg** 翻譯

PREFACE

by Gerald J. Gruman, M.D., Ph.D.

Lake Erie College

前言
Gerald J. Gruman，醫學博士，哲學博士
Lake Erie 學院

PREFACE

by Gerald J. Gruman, M.D., Ph.D.

Lake Erie College

While reading this book, I was reminded of the Belgian businessman who in the early days of World War II heard rumors about the possibility of atomic fission. He ordered a large supply of uranium from the Congo and sent it to warehouses near New York just in time for the atomic bomb project. (On Edgar Sengier, winner of the U.S. Medal of Merit and former president of the Union Miniere du Haut Katanga, see The New York Times, 7-30-63:29.) I must confess that were I interested in business speculation, I should be busily stockpiling equipment needed for Mr. Ettinger's project.

Unlike the creation of the atomic bomb, Mr. Ettinger's proposals are completely benevolent and humanitarian in their intent, so much so that readers may wonder why scientists and physicians are not already applying low-temperature techniques ("cryobiology") to extend human life. To this it must be said that too often there has occurred an unfortunate lag between the scientists' findings in the laboratory and the application of those findings for human welfare.

前言

Gerald J. Gruman，醫學博士，哲學博士
Lake Erie 學院

當我在拜讀這本書的時候，讓我想到了一個比利時的商人，他在第二次世界大戰的初期，聽到了有關原子分裂是有可能的謠言，於是他就從剛果買了一大堆的鈾原料，將之運送到靠近紐約的倉庫，以便能及時趕上作原子彈專案計劃的生意。(請參見紐約時報，7-30-63:29，介紹 Edgar Sengier，美國功勞獎章得主，先前 Miniere du Haut Katanga 聯盟主席)。我必須承認，假使我對生意投資有興趣的話，我現在就應該開始忙於囤積 Ettinger 先生計劃中所需要的設備。

和製造原子彈計劃不同的，Ettinger 先生所建議計劃的意向完全是屬於公益性而且是人道性。但是既然如此，讀者一定會對為何科學家和物理學家到現在還沒有將低溫技術 (冰凍生物學) 應用在人類壽命的延長產生一些疑問。針對這一點我們必須要理解，科學家在實驗室中的發現以及將其發現應用在人類的福祉之間，往往會存在著一個不幸的時間上的延遲。

In 1928, for example, Sir Alexander Fleming discovered that penicillin was remarkably effective in killing germs, but he lacked the capital to prepare sizable quantities of the substance, and nothing was achieved until the massive casualties of World War II stimulated a cooperative search by government and business in Britain and America. By 1944 the drug was performing medical miracles; but what about the lag between 1928 and 1944? No one can calculate the cost of those fifteen years in human suffering. It has been the same with other much-needed innovations: the first anesthetics were suggested in the early 1800's but forty more years of anguish passed before surgical operations became painless, and an even longer struggle was necessary before this benefit was extended to women in childbirth.

Many more illustrations could be given indicating what I think is the most outstanding virtue of Professor Ettinger's book: he is trying to bridge a gap between the world of the research laboratory and that of everyday practice, because he has come upon something which holds great promise for mankind. He has spent years searching the technical literature in a careful and responsible way in order to prepare himself for a vital role: the arousing of general public demand for a new service which science can offer, and the stirring of the conscience of physicians, lawyers, businessmen and government officials so that the demand will be met.

舉個例來說，在 1928 年，**Alexander Fleming** 爵士發現盤尼西林在殺菌上具有很顯著的功效，但是由於他缺乏資金來大量生產盤尼西林，以致於一切都在原地停擺。直到第二次大戰時，因為有大量的傷患產生，才激發起英國和美國政府和商業界的一個聯手的研發。到了 1944 年，此藥才得以展現其神跡式的療效。但是，奈何要有 1928 年到 1944 年間的延宕呢？在此十五年的延宕之間，實在是沒有一個人可以計算出整個人類受苦受難所付出的代價。至於其它一些需求孔急的發明，其情形也都是如此，例如在 1800 年代的初期，就有人第一次提出麻醉學的概念，但是卻要經過四十多年的痛苦等待，外科手術才能夠變成無痛，而且要經過更長時間的掙扎，這種好處才能普及到生產中的婦女。

　　還有許多可以用來說明在 **Ettinger** 教授的著作中，我認為最為卓越超群的優點：他乃是試圖要來填補介於研究實驗世界以及我們日常習性間所存在的一道鴻溝，因為他已經發現了某些對人類有極大潛力的東西。他以一種非常嚴謹並且負責任的態度，在科技文獻中鑽研多年，為的是要準備他自己成為一個關鍵性的角色：喚醒一般大眾對一種科學上可以提供的一種新服務的需求，並且激發醫生，律師，商人和政府官員的良知意識，來讓這樣的需求可以得到滿足。

Mr. Ettinger feels that what he is calling for may happen anyway someday (to some degree, it already is happening), but what he wants to be sure of is that it will happen as soon as possible and in the best possible way. That is why he has adopted a stirring, optimistic writing style, and, in my opinion, he is justified in doing so, because he has a solid grasp of the physical, chemical and biological processes he discusses and a hard-headed appreciation of contemporary technical, economic and social realities.

What is this revolutionary development in science? In brief, it is this: if a man dies today it no longer is appropriate to bury or cremate the body. For there is hope that by keeping it at very low temperatures, physicians of the future may be able to revive him and cure him. And if someone has an "incurable" disease, it is not good practice any more to let him succumb; it is preferable to put the patient into low-temperature storage until better medical facilities become available, or until a cure is discovered. In regard to the scientific and medical bases of this concept, we are fortunate in having the excellent preface by Dr. Rostand who is world-renowned both for his laboratory research and for his understanding of the social and philosophical aspects of science. As Mr. Ettinger states, Dr. Rostand in 1946 was the first to report the protective action of glycerol in the freezing of animal cells.

Ettinger 先生認為他所呼籲的在未來某天一定會成真 (就某個程度而言，其實是已經在實現了)，但是他所追求的乃是要讓其越早實現越好，而且要以最佳的方式實現。這就是為什麼他採取了一種鼓勵和樂觀的書寫筆調。以我個人的觀點來看，他這樣做是完全合理的，因為他已經完整地掌握了他所在探討的主題的物理，化學和生物程序，而且針對當代的科技，經濟和社會上的現實，他也有極為務實的理解。

在科學上有什麼革命性的進展呢? 簡而言之，這就是: 當一個人在今天過世後，如果將其遺體埋葬或是火化，就不再是一種恰當之舉。如果將之以非常低的溫度保存下來，未來仍有復活的希望，因為未來的醫師將有可能使他復甦，然後將他治癒。而且如果有人罹患了一種目前 "無法治癒" 的疾病，讓他持續地惡化下去將不再是一個明智之舉; 最好現在就將此患者以低溫儲存起來，等待更好的醫療科技研發出來，或是研究出一種治癒方法的時候。有關這種新觀念在科學和醫學上的基礎，我們很幸運有 Rostand 博士精采的前言來加以說明。Rostand 博士在其實驗室的研究上，以及其對科學中的社會學和哲學觀點的理解乃是蜚聲國際的。如同 Ettinger 教授所述，Rostand 博士在 1946 年就率先發表甘油在冰凍動物細胞過程中的保護作用。

It also is noteworthy that the English scientist Dr. A. S. Parkes in whose laboratory the glycerol phenomenon independently was rediscovered in 1948 also has spoken favorably about the possibility of cryogenic preservation of the body for indefinite periods of time. (C. E. W. Wolstenholme and M. P. Cameron, eds.: Ciba Foundation colloquia an aging, vol. 1, Boston, 1955: 162-69.)

Mr. Ettinger represents the latest spokesman for a worthy American tradition going back as far as Benjamin Franklin. That eminently practical inventor, philosopher-scientist, and statesman predicted in 1780 that scientific progress would bring about means to lengthen the life span beyond a thousand years. Franklin was delighted with the advances of his time; the lightning rod (his own invention), inoculation for smallpox, the steam engine, flying (manned balloons), etc., and he yearned to see the developments of the future. In a letter to a French scientist, he expressed the wish that he might be awakened in a hundred years to observe America's evolution; the great English surgeon, John Hunter, had a similar idea, hoping to arrange thawing for one year out of every hundred. Franklin also was keenly interested in experiments in resuscitating persons apparently "dead" from drowning or electrocution; in fact, the eighteenth century was fascinated by such activities.

另外值得注意的，**A. S. Parkes** 博士於 **1948** 年在其實驗室中再度獨立地發現了這個甘油的效應，這也正面地說明了利用冰凍保存法來保存人體一段無限定時間的可能性。 (請參見 **C. E. W. Wolstenholme** 和 **M. P. Cameron** , eds.: **Ciba Foundation colloquia an aging, vol. 1, Boston, 1955: 162—69.**)

Ettinger 教授乃是代表了一種可以遙遠追溯到富蘭克林 **(Benjamin Franklin)** 的美國可貴傳統的近代代言人。這個崢嶸頭角的實用發明家，哲學—科學家和政治家，於 **1780** 年就預測科學的進步將可以帶來許多新的方法，來將人類的壽命延長到超過一千年。富蘭克林對在其有生之年中科技的進展已經感到相當地滿意：避雷針(他自己發明的)，牛痘育苗的接種，蒸氣引擎，載人飛行汽球等等，然而他還是引頸企盼看到未來的發展。在其寫給一位法國科學家的信中，他流露出他對能夠在幾百年後能夠被叫醒，以便能夠看看美國進展狀況的願望。英國的一位偉大的外科醫師 **John Hunter** 也有類似的看法，他希望他自己能夠安排每一百年來被解凍一次。富蘭克林對被溺水和觸電而明顯 "死亡" 的人的復甦實驗也是非常地熱衷。其實在十八世紀，當時的人們對這些事情都是非常地好奇和著迷的。

The main pioneers in reviving the "dead" were the Humane Societies set up in Europe and the United States after 1767. (On the Humane Societies, see the article by E. H. Thomson: Bulletin of the History of Medicine, 37:43-51 (1963).) They had to overcome some scorn and ridicule, because, among ignorant and superstitious people, attempts to rescue drowning victims or trapped coal miners were considered utterly foolhardy. But many a conscientious doctor threw himself into the cause, and there were enlightened clergymen to back them up; the Quakers of Philadelphia aided these reforms, and also the great Methodist John Wesley was called into the campaign. An Episcopalian minister concluded in a sermon in 1789 that the Humane Societies deserved his blessing, "Their sole reward is in the holy joy of doing good." As we congratulate ourselves today over the Red Cross and medical successes in artificial respiration, cardiac massage, blood banks and other methods to revive the "dead," we should recognize that Mr. Ettinger is performing the same kind of service and merits our wholehearted support.

Bringing up the question of the nature of death is a major contribution of this hook, and it is one reason why physicians should read it carefully.

使 "死亡" 的人復甦最主要的幾個開拓者乃是於 1767 年之後設立在歐洲和美國的一些人道協會 (Humane Societies)。(有關人道協會請參見文獻: E. H. Thomson: Bulletin of the History of Medicine, 37:43 – 51 (1963).) 他們當時都要承受和克服一些冷嘲和熱諷,因為除了一些無知和迷信的人士之外,試圖要去拯救溺斃或是礦災煤炭工人的舉止都被認為是全然地莽撞。但是還是有一些醫德較高的醫生會投身入這種義舉,而且有一些道行較高的神職人員會全力支援他們。其中有費城的貴格會支援了這些改革,還有衛理會可敬的約翰衛斯禮 (John Wesley) 牧師也被乎召投入這個運動。一個門徒教會的牧師在 1789 年的一場講道中表示,人道協會是應該得到上帝的祝福的 ,"他們靈魂上的獎賞乃在於他們在所作善事中所獲得的神聖喜悅"。現在當我們在慶幸有紅十字會,以及人工呼吸,心臟按摩,血液銀行和其他拯救 "死亡" 者的方法時,我們應該要認知到 Ettinger 教授所在進行的也是一種值得我們全心全意支持的服務和功德。

提出對死亡本質的再思乃是本書的一個重大貢獻,這也是醫師們應該詳加研讀的一個原因。

We tend to accept uncritically as absolute such concepts as "irrevocable damage," "biological death," etc., and we overlook the insidious nature of this "hardening of the categories," (A phrase coined by Dr. Esther Menaker to describe a common "intellectual disease" of professionals and experts.) an intellectual flaw as prevalent and as hampering as hardening of the arteries. This is one of the most useful things about Mr. Ettinger's text; he challenges with admirable tenacity many of these fixed ideas, and every physician will benefit from reading his ingenious attacks on hypotheses we too often take for granted. By serving this function, Mr. Ettinger helps to open original lines of thought and to prevent any lag in the utilization of recent findings in cryobiology, both in practice and in further research.

Of course there are a few points (all peripheral) on which I might not completely agree with Mr. Ettinger; but this has not obscured for me the undeniable logic of his train of thought and the real value of his insight into some of the most difficult problems of modern man. I believe that reviewers and readers in general will find that the core of the book once grasped will never be forgotten and not only will lead to further thought but also to action. We have heard a great deal recently (to our shame) about the costly and childishly sentimental funeral practices referred to as the "American way of death." (Jessica Mitford: The American Way of Death, N.Y., 1963.)

我們一般都趨向於去接受一些非關鍵性的而來當成是絕對的，例如 "不可回復的傷害"，"生物性的死亡"等觀念，而我們都忽視了這種 " 統括的僵化"，中所隱藏的傷害力的本質，(此乃是 Esther Menaker 博士所鑄造出來的一個字句，用來形容一些專業人士和專家所常罹患的一種普遍的 "智慧上的疾病")。這一種智慧上的瑕疵是和血管硬化一樣地普遍，傷害力也是一樣的。這是 Ettinger 教授著作中最犀利的地方，他以無比可敬的耐心挑戰了許多這一類型的頑固思想，藉著閱讀他對一些我們常常會認為理所當然的觀念的睿智攻擊，每一個醫生應該都會有所獲益。藉著淋漓盡致地扮演出他的角色，Ettinger 教授不僅幫我們解放了思想上最原本的路線，而且也防止了在利用冰凍生物學上最先進的發現，不管是在實際應用上或是更進一步的研究上，任何時間上的延宕。

當然在其中的一些論點上 (大都是邊緣性的) 我可能不會全然地同意 Ettinger 教授，然而我並不會因此對於他一系列思想中不可否認的邏輯，以及他對現代人中某些最棘手問題的見解的實質價值而產生任何的混淆。我深信大部分的書評家和讀者，只要掌握了這本書的核心，將永遠不會將其忘記，甚至不僅會誘發出更深層的思考，而且會想要有實質行動的參予。最近很遺憾地，我們有聽到許多有關既昂貴而且幼稚的感性式的葬儀，這被指為是所謂 "美國人的死亡方式" (請參見 Jessica Mitford 的 The American Way of Death. 紐約 1963.)。

Here we have a book which proposes an American way of living on, a demand that our superb (and underemployed) technological facilities be used to implement in a realistic and mature way our avowed belief in the beauty and value of life and health and the immeasurable worth of the individual.

In conclusion, I am reminded of the story about Benjamin Franklin who on one occasion was marvelously rescued from a shipwreck. Having expressed feelings of gratefulness and thanksgiving, he was asked if he intended to build a chapel to memorialize his escape. "No, indeed not," he replied, "I'm going to build a lighthouse!" It is my considered opinion that Mr. Ettinger too has "built a lighthouse," one which throws a powerful light into the years ahead. In the first sudden brightness some persons will be startled, others will ponder curiously the strange, unexpected ways that old perspectives and landmarks have been altered. But those who have faced the pain and the loss and the maddening "absurdity" of human death, whether on a wartime battlefield or in dingy hospital wards - those persons will feel this illumination as a welcome glow of hope in a world which has been waiting so very long.

現在我們有了一本書，它提出一個美國人繼續存活的方法，一個要求將我們所擁有超優的 (然而卻是不常使用的) 科技設施，來實際地而且成熟地應用在能夠反應出我們對美麗，健康和生命價值的堅定信仰，以及個人存在所代表無法衡量的可貴。

　　總而言之，這讓我想到了富蘭克林在一次船難中神奇地被拯救出來的故事。在他表達了他無限的感激和感恩的感受後，他被問及是否有意向要蓋一座教堂來紀念他的大難不死。他回答說，"不，絕對不會。我要蓋的是一座燈塔"！我思慮過後認為 Ettinger 教授也是在 "蓋一座燈塔"，一座朝向未來好幾年投射超高功率亮光的燈塔。在其起初突來的光亮中，有某些人一定會受到驚嚇，而其他人可能對這種老舊景觀和地標被改變的陌生和突然的方法，會開始好奇地思考。但是對那些不管是在煙硝瀰漫的戰場上，或是在污穢絕望的病床側，曾經面對過痛苦和喪失，以及人類死亡的瘋狂和 "荒謬"的人－這些人一定會感受到這道光芒乃是一道整個世界已經期盼良久，如大旱雲霓般的希望之光。

CHAPTER I

Frozen Death, Frozen Sleep, and Some Consequences

死亡冰凍，睡眠冰凍，及其一些後果

CHAPTER I

Frozen Death, Frozen Sleep, and Some Consequences

Most of us now living have a chance for personal, physical immortality.

This remarkable proposition - which may soon become a pivot of personal and national life - is easily understood by joining one established fact to one reasonable assumption.

The fact: At very low temperatures it is possible, *right now*, to preserve dead people with essentially no deterioration, indefinitely. (Details and references will be supplied.)

The assumption: If civilization endures, medical science should *eventually* be able to repair almost any damage to the human body, including freezing damage and senile debility or other cause of death. (Definite reasons for such optimism will be given.)

Hence we need only arrange to have our bodies, *after we die*, stored in suitable freezers against the time when science may be able to help us. No matter what kills us, whether old age or disease, and even if freezing techniques are still crude when we die, *sooner or later* our friends of the future should be equal to the task of reviving and curing us. This is the essence of the main argument.

第一章

死亡冰凍，睡眠冰凍，及其一些後果

我們現在活著的人，大多數都有機會得到個人肉體上的永生不死。

這個如此突顯的提議——其可能很快地就會變成個人和整個國家生活上的一個樞軸——藉著將一個既成的事實和一個合理的假設結合在一起，就可以很容易地被理解。

既成的事實: *目前*，在非常低的溫度之下，是有可能將一個死人無限期地保存，而幾乎不會產生任何的變質。(後面將提供一些細節和參考資料)

合理的假設: 假使文明可以持續發展，*最終地*，醫療科學幾乎將有可能修復人體上任何的傷害，其中包括冰凍傷害和老化疾病或是其他導致死亡的原因。(後面將提供一些能夠如此樂觀的確切理由。)

因此*在我們死後*，我們只要將我們的身體安排去儲存在一個合適的冰凍櫃中，來對抗當科學有辦法幫助我們時的時間。不管我們死亡的原因是什麼，或許是老化或許是疾病，甚或是當我們死亡時的冰凍技術還是非常粗糙，*早晚總有一天*，我們未來的朋友應該會有能力將我們復甦並且治癒我們。這就是整個主要論點中的精闢之處。

The arrangements will no doubt be handled at first by individuals, then by private companies, and perhaps later by the Social Security system.

By preserving our bodies in as nearly life-like a condition as possible, it is clear that you and I, *right now*, have a chance to avoid permanent death. But is it a substantial chance, or only a remote one? I believe the odds are excitingly favorable, and it is the purpose of this hook to make this belief plausible. If it is made plausible, the necessary efforts will be encouraged further to improve the odds.

It is my hope that the cumulative weight of the discussion will convince the reader that his own life is at stake, and those of his family, and that his personal efforts are urgently needed in this mighty undertaking. (The pun should be forgivable; it is impossible consistently to accord the subject the awesome dignity it deserves.)

Suspended Life and Suspended Death

It must be made very clear that our basic program is *not* one of "suspended animation," and does *not* depend on any special timetable of scientific progress, but can be instituted *immediately*. To make sure of our orientation, let us review the meaning of suspended animation and of the several kinds of death.

在剛開始的時候,這些安排無疑地將會是由一些個人來經手,之後則會由一些私人公司來處理,到最後有可能會變成由社會保險系統來統籌安排。

藉著將我們的身體保存在越接近活的狀態越好的情形下,很清楚地你我,*現在*,都有一個機會來避開永久的死亡。但是到底這會是一個實質的機會,還是僅會是一個渺茫的呢**?**我相信其達成的機率是高到令人興奮,而撰寫這本書的目的就是要來證明這種信念是有根有據可信的。假使在證明其是可信之後,剩下所需的努力將是百尺竿頭更上一層樓地來增高其機率。

我希望藉由討論所累積的力量,將足夠來說服所有的讀者,讓大家能夠感受到所在賭注不僅是自己的生命,連其家人的生命亦然,因此在這項偉大的舉動中,他個人的一些努力是需求孔急的。(請原諒其中的文字遊戲:因為要能夠一直都一致地維持此主題所該獲得的無比尊嚴是不可能的。)

生命暫眠和死亡暫眠

大家必須很清楚地知道,我們的基本計劃*不是*屬於所謂 "活體暫眠" 的一種,並且不仰賴著科技進步中的任何一個特殊的時間進度表,而且是可以馬上著手去建構而成的。為了要確定我們發展的方向,讓我們一起來重新檢視活體暫眠以及許多不同類型死亡的意義。

Suspended animation refers to a standstill in the life processes of the body. It is a stasis that can be imposed and removed at will, and the subject is regarded as alive at all times. In some simple life forms suspended animation can be produced simply by drying, and reanimation by moistening them again; in fact, certain bacteria found embedded in salt have been reported revived after hundreds of millions of years. For humans, the only likely way to induce suspended animation is by freezing, but full recovery after complete freezing has not yet been achieved with any mammal.

The subtle distinction between life and death is evident in the case of the dried bacteria, which were regarded as alive merely because they were potentially capable of displaying life processes. In fact, we recognize at least five kinds of death, which must be kept firmly in mind.

"Clinical death" is the kind we most frequently have in mind, its criteria being cessation of heartbeat and breathing.

"Biological death" has been defined by Dr. A. Parkes as the state from which resuscitation of the body as a whole is impossible by currently known means. This is very logical, but also very odd: a frozen body might lie around for years in a "dead" condition, then all at once come alive, without any physical change whatever, as soon as someone found a means of resuscitation.

"Cellular death" refers to irreversible degeneration of the individual tiny cells of our bodies.

活體暫眠乃是指人體生命過程中的一個原地停頓。此乃是一種可以隨意加上或是解除的靜止狀態，而當事者乃是被認為一直都是活著的。在某些簡單形式的生命中，活體暫眠可以藉著乾燥就可以很容易地產生，而且藉著再度的潤濕它們，就可以使其復甦過來。事實上，某些被深埋在鹽層的細菌，科學研究發現在經過好幾億年之後，還可以使它們復活起來。然而就人類而言，要誘發活體暫眠唯一的方法似乎只能藉著冰凍，而且沒有任何一種的哺乳類動物，在經過完全的冰凍之後，曾經被成功地完全的復原。

　　介於生命和死亡間的微妙區別，可以從被乾燥的細菌上看得很清楚。它們之所以被認定為還活著，僅因為它們還具有展現生命的潛在能力。事實上我們有找出至少五種所謂的死亡，這是我們必須牢記在心的。

　　"醫療上的死亡"乃是我們經驗中最常見的一種，其認定標準乃是心跳和呼吸的停止。

　　"生物性的死亡"乃是由 A. Parkes 博士所定義出來的。此乃是藉由目前所有已知的方法，仍然無法將人體救活的一種狀況。這雖然是非常合乎邏輯，然而也是非常地矛盾：一個被冰凍的身體可能以一種"死亡"的狀態躺在那裡好幾個年頭，然後當有人發現了使其獲救的方法，就可以在沒有任何現狀改變之下，突然間使其復活過來。

　　"細胞上的死亡"乃是指我們身體中那些單一微小細胞中的一種不可逆的解體。

The questions of legal death and religious death will be left for later chapters.

The important point is that a man does not go like the one-horse shay, but dies little by little usually, in imperceptible gradations, and the question of reversibility at any stage depends on the state of medical art. Clinical death is often reversible; the criteria of biological death are constantly changing; and even cellular death is a matter of degree, since it is possible for an individual cell to be made nonfunctional by minor and eventually reparable damage.

Suspended death, then, will refer to the condition of a biologically dead body which has been frozen and stored at a very low temperature, so that degeneration is arrested and not progressive. The body can be thought of as dead, but not very dead; it cannot be revived by present methods, but the condition of most of the cells may not differ too greatly from that in life.

There is also an interesting intermediate condition between suspended life and suspended death, which will be mentioned in a later chapter.

Future and Present Options

When full-fledged suspended animation becomes practicable, a wide range of options will be available. For example, the feeble aged and the incurably ill may choose to suspend life and await a day when cures are known.

有關 "法律上的死亡" 和 "宗教上的死亡" 的問題則保留到下面的章節再加以探討。

要點是一個人的死亡不會像是一台單匹馬車一樣,一馬垮就全垮,通常都會是以一種難察秋毫的速度逐漸局部地死亡,而其在任何一個死亡階段的可逆性,則完全是取決於醫療技術的境界。醫療上的死亡往往是可逆轉的。生物性死亡的認定標準則是一直在改變。而甚至於細胞上的死亡也是一種相對程度上的事情,因為單一細胞之所以會變成完全沒有活性,開始都是由一些微小成因的累積,而造成最終不可回天的傷害。

於是,死亡暫眠所指的乃是一個生物性死亡的人體,被冰凍且儲存在一個非常低溫度的狀態,以致於其解體現象完全停止而不繼續進行。雖然這個人體是可以被判定為已經死亡,然而卻不是絕望的死亡。雖然以目前的醫療技術無法使其復活,但是其中大部分細胞的狀況,比起其活著的時候並沒有改變多少。

介於生命暫眠和死亡暫眠之間,還存在著一種饒趣的過度狀況,容我們於下面的章節中再加以討論。

未來和現在的一些選擇

未來當整套的活體暫眠技術變成可行時,我們將會有一個大範圍的選擇可能。例如,體虛孱弱的老人和無藥可救的病人就可以選擇來暫停其生命,等待醫療方法被開發出來的那一天。

On the other hand, many people may still choose to be frozen only after natural death - but the techniques of suspended animation, applied after clinical death but before biological death, should ensure that their condition is still one of suspended life. (It is not self-evident that techniques applicable to a living person are also suitable for one clinically dead, but reasons for thinking so will be produced later.)

The chief value of research on suspended animation, then, is that it will develop new freezing techniques, ways to avoid freezing damage. When this is achieved, we will be able to preserve our freshly dead bodies with only the damage of old age or disease, and without the additional insult of damage by crude freezing methods, and thus our chances of early resuscitation will be vastly improved.

(How strange that the many popular articles on suspended animation have mentioned chiefly its possible use by astronauts on long interstellar voyages! This aspect is trivial. Its importance lies not in travel to the stars, for the few, but in travel to the future, for the many. It will open a veritable "door into summer" for all of us.)

Research in freezing techniques is proceeding actively, although so far on a relatively small scale, at a number of laboratories and hospitals in the United States, France, Britain, Russia, and elsewhere. Some small animals, and some types of human tissue, have been deep-frozen and successfully restored to life.

另一方面，有很多人還是會選擇等到自然死亡之後才進行冰凍，這也無傷大雅，因為如果將活體暫眠的技術應用在醫療死亡之後，但是是在生物死亡之前的話，應該可以保證他們還是屬於一種生命暫眠的狀態 (這種應用在活人的技術也可以應用在醫療死亡的人上，其原理不是很顯然易懂，因此在後面的章節中，我們將會特別闡述其可行的理由。)

　　因此，研究活體暫眠技術的主要價值乃是在開發出新的冰凍技術，尋找出能夠避免冰凍傷害的方法。當其目標達成時，我們就可以完整地保存剛剛死亡者的身體，其傷害將只有其本來的老化或是疾病，而沒有任何因為酷冷的冰凍技術之攻擊所造成的額外的傷害。我們能夠早日復甦的機率也因而可以大大地提昇。

　　(有關活體暫眠的許多著名的文獻，主要地都是在討論其可能應用在在長途星際旅行中的太空人上，這實在是有點奇怪！這方面的應用其實是有點次要。其應用重點應該不是在於那些要旅行到其他星球的少數人，而應該是在想要延伸生命到未來的多數人。這技術應該會為我們所有的人開啟一扇真正 "進入盛夏的大門"。)

　　冰凍技術的研究目前雖然都屬於相對性較小的規模，但是在許多美國，法國，英國，俄國和其他一些地方的實驗室和醫院，都有在積極地進行中。一些較小的動物，以及某些型態的人體組織，都曾經被深度地冰凍而成功地將其恢復生命過。

Actual full-body freezing and suspended animation of a human being is anticipated fairly soon by some workers. Dr. James F. Connell, Jr. (St. Vincent's Hospital, New York) is reported in 1962 to have said, "If all the medical personnel involved with this problem make a concerted effort, we will do it in less than five years."

Research work will be multiplied and accelerated if sufficient demand appears for freezer programs. Should this happen, most of us now living will have the benefit of freezing by advanced techniques, so that our bodies will be preserved in much better condition than is now possible.

If feasible, therefore, one should contrive to stay alive for the next few years, since the odds will improve rapidly during this time.

For the present, we must rely on the basic program of suspended death. It is simply proposed that, after one dies a natural death, his body be frozen and preserved at a very low temperature - perhaps near absolute zero, the lowest possible temperature - which will prevent further deterioration for an indefinite period. The body will be damaged by disease or old age which is the cause of death, and will be further damaged (although in some eases probably not much, as we shall see) by our current freezing methods.

某些研究科學家預期，一個人真正全身冰凍和活體暫眠的可能不久就會達成。紐約聖文森醫院的 James F. Connell, Jr. 博士在 1962 年曾經發表過，"假使所有參予這個問題的醫學研究人員能夠意志集中力量集中的話，我們在少於五年的時間內就可以達成目標。"

　　對冰凍計劃假使能夠有足夠的需求顯現的話，此方面的研究工作將會有倍數的增加，其進展也會加速。這狀況如果成真，現在活著的人大多數將可能享受到先進技術冰凍法的好處，這樣我們的身體將可以被以目前技術好上許多倍的狀態保存下來。

　　因此，假使可能的話，每一個人都應該努力來渡過最近的幾年，因為在這段時間內，成功的勝算將會急速地增高。

　　在目前的狀況下，我們不得不只能仰賴基本的死亡暫眠的計劃。此計劃只是建議當一個人在自然死亡之後，他的身體被冰凍起來，並且保存在一個非常低的溫度，大約是接近絕對溫度零度，也就是最低可能的溫度下，這樣就可以在一段不定期的時間內，防止其更進一步的腐朽。這個身體一定會有來自疾病或是老化的傷害，因為這乃是其死亡的原因，但是也會有來自我們目前冰凍方法的進一步傷害（雖然在某些案例中顯示出此傷害可能不多，於後面章節將有所討論）。

But it will not decay or suffer any more changes, and one assumes that at some date scientists will be able to restore life, health, and vigor - and these, in fact, in greater measure than was ever enjoyed in the first life. (This is a tall order, of course, and one of the chief aims of this book is to make it seem reasonable.)

After a Moment of Sleep

The tired old man, then, will close his eyes, and he can think of his impending temporary death as another period under anesthesia in the hospital. Centuries may pass, but to him there will be only a moment of sleep without dreams.

After awakening, he may already be again young and virile, having been rejuvenated while unconscious; or he may be gradually renovated through treatment after awakening. In any case, he will have the physique of a Charles Atlas if he wants it, and his weary and faded wife, if she chooses, may rival Miss Universe. Much more important, they will be gradually improved in mentality and personality. They will not find themselves idiot strangers in a lonely and baffling world, but will be made fully educable and integrated.

If civilization endures, if the Golden Age materializes, the future will reveal a wonderful world indeed, a vista to excite the mind and thrill the heart. It will be bigger and better than the present - but not only that.

但是畢竟他將不再會有進一步的腐化，或是必需去承受其它任何更多的變化，因此我們可以假設到某個時日，科學家們將有辦法來恢復其生命，健康和活力。而且事實上，他將可以享受比其第一條生命時更多的生命，健康和活力。(這當然是一個極高度的要求，但是使其看起來合情合理也是寫這本書的目的之一。)

在一段睡眠之後

這個身心疲倦的老人接著將可以閉上他的雙眼，把他將要接受的暫時死亡，想像成他在醫院中的另一段被麻醉的時段。任時光可能如白駒過隙般地經過好幾百年，但是對他而言，這可能僅是一小段沒有作夢的睡眠。

因為在無意識當中他已經被回春了，所以在被喚醒之後，他可能已經再度變成既年輕又有幹勁。或是他也可以在被喚醒之後才來接受治療，進行逐步的更新。不管如何，假使他想要的話，他將可以擁有 Charles Atlas (希臘神話中用肩膀撐天的天神) 的身段。而其已經黃花枯凋的老伴，假使她有意的話，也可以變成比環球小姐還要美的尤物。更重要的是，他們的想法和個性將會逐漸的改善。他們將不再會覺得自己像是活在一個孤單而且怪異世界中的愚蠢的陌生人，反而將會變成非常地可受教和完全地投入。

假使人類文明可以持續下去的話，假使這個黃金時代來臨的話，未來的確將會彰顯出一個美妙的世界，一個可以興奮心靈激盪魂魄的景象。未來將會是比當今更壯闊更美好—但也不僅是如此而已。

It will not be just the present, king-sized and chocolate covered; it will be different. The key difference will be in people; we will remold, nearer to the heart's desire, not just the world, but ourselves as well. And "ourselves" refers to people, not just posterity. You and I, the frozen, the resuscitated, will be not merely revived and cured, but enlarged and improved, made fit to work, play, and perhaps fight, on a grand scale and in a grand style. Specific reasons for such expectations will be presented.

Clearly, the freezer is more attractive than the grave, even if one has doubts about the future capabilities of science. With bad luck, the frozen people will simply remain dead, as they would have in the grave. But with good luck, the manifest destiny of science will be realized, and the resuscitated will drink the wine of centuries unborn. The likely prize is so enormous that even slender odds would be worth embracing.

Problems and Side Effects

In order to remove the prospect of immortality from the realm of thin, hazy speculation or daydreams and secure it in the domain of emotional conviction and work-a-day policy, it is essential that the discussion assumes some scope and provides some background detail. The gist of the main argument has already been given, but it needs to be filled out and buttressed.

其所帶來的將不僅是如塗滿巧克力的大包禮物；它將會是一份全然不同的禮物。其中最大的不同將會是發生在人上面；不光是外在的世界，我們自己也是一樣，將可以被重新塑造到較接近我們自己心裡面所想要的形象。而其中所指的“我們自己”乃是指所有的人，而不是光指我們的子孫。你和我，這些被冰凍而再復活的人，將不再僅是會被復甦和被治癒，而是會被擴增和被改良，使我們更適宜在一個更高級的水準上和一個更優雅的風度上工作，遊戲，甚或是打鬥。至於能期望這些的相關特定理由，容我們於後面章節中再加以敘述。

甚至於對那些懷疑未來科學能耐的人而言，很顯然地，冰凍櫃總會比墳墓更具有吸引力。如果運氣不好的話，這些冰凍的人將只會像其被埋在墳墓中一樣，僅僅維持其死亡的狀態而已。但是如果運氣好的話，萬一這些科學中所彰顯的未來命運得以成真的話，那麼這些因此而得以復活的人，將可以暢飲幾世紀以來未曾有過的葡萄美酒。這個可能的獎賞是如此的巨大，因此就算其勝算是極其渺茫，也值得我們熱情地去擁抱它。

一些問題和副作用

為了要讓永生的盼望跳脫出一些荒誕怪異虛無飄渺的臆測或是白日夢的範疇，而要使其在心理感性的信念和日常例行的政策的領域中佔有一席穩固之地，因此在討論中一定要涵蓋某一特定範圍，並且要提供一些和其相關的背景細節。雖然主要辯論的核心重點已經被訂定了，但是它還需要被填補和被支撐。

Many obvious objections must be met, a host of troublesome questions answered.

How much progress in freezing techniques has actually been made? How much is known about freezing damage? How severe is the damage produced by current methods of freezing, and what reasons, other than vague optimism, are there for thinking the damage may be reversible? Can frostbite be cured?

Since the brain usually begins to deteriorate within a few minutes after breathing stops, how will it be possible to freeze the body soon enough? Considering the varied circumstances of death, how can one cope with the diverse practical problems that will be faced by the pioneers in treating and storing bodies?

Do you have a legal right to freeze a relative? Will failure-to freeze be considered murder or negligent homicide? Will there be an increase in mercy killings and suicides? Can a corpse have legal rights and obligations? Can a corpse vote?

Can families be kept together? Will widowers and widows be allowed to marry again in the first life? What will happen to the resuscitated person confronted with two or more ex-husbands or wives? Is there a conflict between the freezer program and religion, or should the freezers be considered merely the latest in a long series of medical efforts to save and prolong life?

有許多明顯的反對力量必須要去對抗，也還有一序列棘手的問題必須要去解答。

到現在在冰凍技術上到底已經有了多少實質的進展呢？對冰凍傷害理解的程度到底是如何呢？以現有的冰凍技術所產生的冰凍傷害到底是會有多嚴重呢？其產生的原因是什麼呢？除了模糊的樂觀主義之外，是否有去深思傷害可以被逆轉的可能性呢？凍瘡是否可以被治癒呢？

既然在呼吸停止後的短暫幾分鐘內，人腦通常就會開始腐壞，如何才有可能及時地將身體冰凍呢？顧及死亡的各種不同的狀態，我們如何來克服在處理和儲存這些先鋒部隊時，所可能面對的各式各樣現實上的問題呢？

你是否有法律上的權利來冰凍你的親戚呢？冰凍失敗是否被認定為蓄意謀殺或是過失殺人呢？安樂死和自殺案例是否會因此而大幅增加呢？一具保存的屍體是否具有法律上的權利和義務呢？一具屍體是否享有投票權呢？

家族成員可否被儲存在一起呢？在第一次生命中的鰥夫和寡婦是否容許他們再度結婚呢？當一個復活的人必須面對兩個以上的前夫或是前妻時，該怎麼辦呢？冰凍計劃和宗教之間是否會有衝突呢？或是這些冰凍櫃僅應該被認定為是用來拯救和延長生命中，所應用一長序列醫療動作中的最後階段呢？

If a Christian refuses a chance at extended life through freezing, does this amount to suicide?

Will the cost of dying become so high that we cannot afford it? If we freeze every American, the current population alone will produce something like fifteen million tons of bodies; where's all the money come from, and where can we stack them all?

What about the population problem? When the frozen are revived, where will the throngs of ancestors find lebensraum? Do we have a right to impose ourselves on our descendants, like a mob of poor relations come to dinner? Who needs us? Will it be only the selfish and cowardly who are frozen?

Even if the future welcomes us and makes room for us, will we like it? Even if we like it at first, will we not become bored? How can a mere human endure, let alone enjoy, thousands of years of life? And if we cease being human and become superhuman, will we still be ourselves? How much can a man change without losing his essence?

In fact, some of the most profound questions of philosophy are forced to the level of practical affairs. What is a man? What is death? What is the purpose of life?

假使一個基督徒拒絕接受通過冰凍計劃所賦予的一個延長生命的機會，這是否會被認為是一種自殺的行為呢？

　　死亡過程的成本是否會變得太高，以致於我們都負擔不起呢？假使我們要去冰凍美一個美國人，光就目前的人口而言，就可能產出一千五百萬噸的軀體，這些經費要從哪裡來，而且哪裡可以全部堆放他們呢？

　人口問題該怎麼辦呢？假使所有的冰凍人都復活了，這一大群祖先們要去哪裡找到居住空間呢？我們是否有權利來強把自己加諸在後代的子孫上面，好像乞丐趕廟公一樣呢？有誰會需要我們呢？是否僅有那些自私者和懦夫會選擇冰凍呢？

　　就算是未來的世界會歡迎我們，並且願意容納接受我們，我們自己是否會喜歡呢？甚至就算我們一剛開始還蠻喜歡的，但是久了之後是否會覺得無聊呢？甭談去享受了，一個人如何忍受幾千年的生命呢？還有假使我們不再是一般人，而是可以變成超級人，我們是否還是自己呢？一個人到底可以改變多少，而還可以不失去他的本質呢？

　　事實上有許多哲學上非常深邃的問題，都必須要去思考一些現實上的事務。何謂人呢？何謂死亡呢？什麼是生命的意義呢？

How will the answers to these questions affect existing problems? Will a freezer program cause sharper competition or more cooperation among individuals and nations? Will a nuclear war become more likely or less? Will a man looking forward to thousands of years be less inclined to rock the boat and more inclined to practice the Golden Rule?

An attempt will be made to throw some light into all these dark corners.

這些問題的解答將會如何來影響現存的問題呢？一個冰凍計劃是否會在人與人之間和國家之間，引發更尖銳的競爭或是更密切的合作呢？核子戰爭將會變得更有可能或是更不可能呢？一個可以有幾千年壽命的人，他會較喜歡去冒險進取還是會較傾向於去明哲保身呢？

　　我們將會試著去尋求一些亮光來照亮這些未知的暗處。

CHAPTER II

The Effects of Freezing and Cooling

冰凍和冷卻的一些效應

CHAPTER II

The Effects of Freezing and Cooling

If you are about forty years old now, then probably when you die, in another thirty or forty years, physicians or technicians paid by your insurance company will bank your blood, perfuse your parts, and lay you to rest – not eternal rest, but temporary, and not in the cold ground, but in a much colder freezer. A few years later, perhaps they will slide your wife in beside you.

At first thought, many people find this notion both implausible and a little repellent. They may find it repellent because their minds associate a freezer with dead meat. They find it implausible, because they know a lamb chop looks pretty inert to begin with, and furthermore begins to spoil after a very few years in a freezer at 0 degrees Fahrenheit.

It is also recalled that we sometimes have to chop off a severely frostbitten toe; we cannot revive it, even though the rest of the body is alive. How, then, can we hope to revive a man frozen throughout his very vitals? How can we have any confidence that it will ever be possible?

A mere, generalized optimism is certainly not convincing.

第二章

冰凍和冷卻的一些效應

假使你現在大約已經四十歲,那麼很有可能再經過另外一個三四十年,當你死的時候,就會有保險公司所聘僱的醫生或是技術人員,來將你的血液存入血庫,擴張你的肢體,然後將你送入安息—是暫時的而非永遠的安息,而且不是在陰冷的地下,而是在一個要冷上許多的冰凍庫。經過幾年之後,或許他們也會將你的老婆送來和你作伴。

許多人對這個觀念的第一個感覺,都會感到非常地難以置信,而且都會有一點點的排斥。他們之所以會有排拒感,乃是因為他們的心理上都將冰凍庫和死豬肉聯想在一起了。他們之所以會感到難以置信,乃是因為他們知道一塊羊排儲存在華氏零度的冰凍庫時,一剛開始看起來都是非常地穩定,可是經過沒有幾年之後就會開始腐壞。

大家也會想到,我們有時候都必須要去切除一根被嚴重凍瘡的腳趾,雖然他身體的其它部分都是好好的,但是我們就是沒辦法來使其痊癒。既然如此,我們怎麼可能會奢想要去使一個全身重要器官都凍透的人復活呢? 我們如何有信心來認為這種未來將會有可能呢?

光是靠著一種以偏蓋全的樂觀主義,絕對是無法令人信服的。

It is all very well to say that future science will surpass imagination; but will it be able to take a tub of frozen corned beef hash, and from this reconstitute a steer - the same steer that went into the hash? We are interested in something that is probable, and not just barely conceivable. If our chances were no better than those of the hypothetical steer, we would not want to be bothered.

To provide a basis for reasonable confidence, let us examine carefully some of the salient facts and estimates concerning the effects on living animals of cooling and freezing.

Long-term Storage

Our basic argument was based on one fact and one assumption. The fact -that it is possible, right now, to preserve dead people with essentially no deterioration, indefinitely - is easily established.

It is a well-known principle of chemistry that at temperatures near absolute zero (about -273C or -459F) reaction rates generally become negligibly small. The molecules are nearly motionless. The life processes of any organism cooled near this extreme should become immeasurably slow, and also any processes of decay. Actual observation confirms this theoretical principle.

我們都可以大言不慚地說未來的科學一定會超過我們所能想像的，但是科學是否將有辦法從一桶冰凍的醃牛肉末中，重組出一隻完整的牛，也就是原來那一隻變成肉末的牛呢？我們有興趣的乃是一些有可能的事，而不是一些僅是憑空夢想的。假使我們成功的機會沒有比此假設中的牛的機會要大過許多，我們就大可不必要去庸人自擾了。

為了要找到合理信心的根據，針對有關動物活體冷卻和冰凍的一些突顯的事實和預估，讓我們一起來小心地檢視。

長期的儲存

我們的基本論點乃是架構於一個事實和一個假設。現在我們可以在幾乎沒有任何腐化的情形之下，來無限期地保存死亡的人，這個事實乃是很容易就可以被建立的。

當溫度在接近絕對零度時 (大約是攝氏零下273度或是華氏零下459度)，任何反應速率一般都會變成小到可以被忽略，這乃是一個眾所皆知的化學原理。所有的分子幾乎都是不動的。任何生物被冰凍到這個極致時，其生命的活動都應該會變成無法量測地慢，任何腐敗的過程也會是一樣。而在實際的觀察中也證實了這個理論性的原理。

Dr. Harold T. Meryman (Naval Medical Research Institute, National Naval Medical Center, Bethesda, Maryland), a leading authority in the field, says, "Under any circumstances, storage in liquid nitrogen, at - 197C can be considered as essentially indefinite." (68)

Dr. Humberto Fernandez-Moran (University of Chicago), a prominent expert in biophysics, notes that ". . . no detectable metabolic activity has been reported at liquid nitrogen temperatures . . ." He points out, however, that activity involving short-lived molecular fragments called "free radicals" can occur at - 197C and that long-term storage should perhaps be at liquid helium temperatures, namely within a few degrees of absolute zero. The reaction rates at liquid helium temperatures are calculated to be slower than at liquid nitrogen temperatures by a factor of about ten trillion! (30)

Many other investigators have written to the same effect. The consensus of the best-informed opinion, based on long observation as well as theory, indicates that a body cooled by liquid nitrogen can be stored without significant changes or deterioration for a period measured at least in years and probably in centuries. A body cooled by liquid helium will keep, for all practical purposes, forever.

Clearly, then, the storage problem is not the main difficulty. Whatever condition the body is in when it reaches the storage temperature, that is the condition in which it will remain for as long as it is necessary to keep it.

美國馬立蘭州伯賽大國家海軍醫學中心海軍醫學研究所的 Harold T. Meryman 博士，他是在這個領域裡的一個領先的權威，他表示 "在任何情況下，只要是在液態氮中的儲存，在攝氏零下 197 度中，實質上都可以被認定為是無限期的。" (68)

美國芝加哥大學的 Humberto Fernandez-Morgan 博士，他是生物物理學的一個傑出的專家，他強調 ".......在液態氮的溫度下，科學界沒有發現任何可以偵測到的新陳代謝的活動........"。 然而他指出，有關所謂 "自由基" 的短暫生命分子片段的活動，在攝氏零下 197 度時還是有可能發生，因此長期的儲存可能必須要在液態氦的溫度下，也就是要在接近絕對零度的幾度之間。在液態氦溫度下所計算出來的反應速率，比起液態氮溫度的反應速率要慢上十兆倍以上! (30)

許多其他的科學研究人士也發表了同樣的結果。這些建立在長期的科學觀察以及科學理論上，由有最佳資訊的意見中所產生的共識，在在都指出一個身體在以液態氮來冷卻之後，就可以儲存到一段至少以幾年或可能是幾百年來計量的時間，而不會產生任何顯著的變化或是腐敗。而如果以液態氦來冷卻，一個身體在各種實用的目的上，則將可能永遠地保存。

因此很明顯地，儲存的問題並非是主要的瓶頸。不管人體當達到儲存溫度時的狀況是如何，此狀況將可以依保存時間上的需要，而一直保持下去。

If it is alive, it will remain alive; if it is somewhat damaged, it will remain somewhat damaged.

The principal hazards pertain to the freezing and thawing processes. Let us next inquire what progress has been made in actually freezing specimens and restoring them to active life.

Successes in Freezing Animals and Tissues

Among smaller and lower organisms, there are many which can survive actual hard freezing at temperatures far below the freezing point, even without any special protection, and others which can be assisted to do so.

Becquerel has found that certain minute, primitive animals, which can tolerate dehydration, can be cooled, after drying, to within a fraction of a degree of absolute zero, and after rewarming and remoistening revive fully. (5) Since the water had been removed before freezing, there had been no damage from ice crystals.

Two Japanese scientists, Asahina and Aoki, worked with larvae of a certain insect, Cnidocampa flavescens. The larvae were removed from their cocoons, kept for one day at -30C, and then immersed in liquid oxygen at-180C. After thawing, their hearts resumed beating, and some of them lived to their next developmental stage, that of "imago," but none completed metamorphosis to the adult stage. (2)

假使當時是活的，則將會一直保持活的狀態。假使當時是有點受損的，則將會一直維持有點受損的狀況。

最主要的危險乃是和冰凍和解凍的過程有關。下面就讓我們一起來探討在一些樣本實際的冰凍上和將其恢復到活生生的生命上，科學界到底有些什麼樣的進展。

冰凍動物和組織的成功案例

在較微小和較低等的生物中，有許多的生物都可以在低於冰點許多的溫度下，經歷過實質強烈的冰凍而存活下來，其中有些甚至於都不需要任何特別的保護就可以，而有些則只要一些幫助就可以。

Becquerel 發現，有某些會耐脫水的微小原始動物，在經過乾燥之後，可以被冰凍到接近絕對零度幾分之幾度下，然後經過再加溫和再加濕後，而被完全地恢復其生命。(5) 因為在冰凍之前水已經被移除了，所以其中就不會有由冰晶所產生的傷害。

有兩位日本的科學家，Asahina 和Aoki，他們研究了某一種稱為Cnidocampa flavescens昆蟲的幼蟲。這些幼蟲首先被從其蟲蛹中取出來，在攝氏零下30度保持一天，然後再將其浸泡在攝氏零下180度的液態氧中。在經過解凍之後，它們的心臟都能夠恢復跳動，而且其中某些還能夠活到下一個成長階段，也就是所謂的 "成蟲，imago"，但是其中就是沒有一隻可以完成其整個蛻變程序而達到成熟昆蟲的階段。(2)

It was thought that the pre-freezing period of one day at -30C allowed growth of ice crystals outside, rather than inside, the cells; that is, the ice crystals formed in the intercellular spaces.

Many protective agents have been tried to reduce damage to animal tissues in freezing; perhaps the most successful of these has been glycerol. Professor Jean Rostand, working with frog spermatozoa, provided the first evidence; motility of the sperm was preserved for several days at -4C to -6C. (94) (The freezing point of pure water at standard pressure is 0C.) Subsequently it has been found that certain cold-hardy insects naturally contain glycerol in their bodies! (110)

Another protective agent sometimes used successfully is ethylene glycol, a solution of which was used by Dr. B. J. Luyet and Dr. M. C. Hartring in freezing vinegar eels, anguillula aceti. The eels survived immersion in liquid air at about - 190C, provided both cooling and rewarming were rapid. (110) It was thought that the ethylene glycol caused dehydration, and induced a vitreous rather than a crystalline condition of the water in the cells.

Clams on certain northern shores, exposed to temperatures far below 0 degrees celsius when the tide runs out, apparently become solidly frozen and thawed twice daily for weeks on end, yet survive. It is suspected that these organisms also may secrete a natural protective agent of some kind, and investigation is continuing. (110)

科學家認為這乃是因為在攝氏零下 30 度冷卻一天的預先冰凍時段，可以讓冰晶在細胞外而不在細胞內長成，也就是讓冰晶形成在介於細胞間的空隙。

科學家也試驗了許多在冰凍時可以減少動物組織傷害的保護製劑，而其中最有效的可能就非甘油莫屬了。**Jean Rostand** 教授在其針對青蛙精子的研究工作成果，提供了相關的第一個證據。在攝氏零下 4 度到零下 6 度之間，精子的活動性可以被保留很多天。**(94)** (在標準一大氣壓之下，純水的冰點是攝氏零度。) 後續的科學研究發現某些能夠抗低溫的昆蟲，其體內都自然地含有甘油! **(110)**

另外一種有些時候也極靈光的保護製劑就是乙二醇， 此乃是 **B. J. Luyet** 博士和 **M. C. Hartring** 博士用來冰凍醋蛆 和醋鰻時所使用的溶液。假使在冰凍和重溫的過程都能快速完成的話，這些鰻魚都能夠歷經液態空氣攝氏零下 190 度的浸泡而存活下來。 **(110)** 科學家認為此乃是因為乙二醇可以導致脫水，而誘發其細胞中的水分產生一種玻璃化而非是一種微晶化的狀態。

在某些北方海岸的蚌類動物，當退潮的時候都要暴露在大大低於攝氏零度的溫度。它們很顯然地在連續好幾個星期中，每天都要被完全地冰凍和解凍兩次，然而它們還是能夠活下來。科學家臆測這些生物也是可能會分泌某一種類的保護成分，他們仍繼續在對其進一步深入研究當中。 **(110)**

When we turn our attention to larger and more highly developed forms of life, we find there have been many successes in freezing and reviving cells, tissues, and even organs. Usually protective agents have been required, but not in all cases.

Bull semen has been treated with glycerol, stored at -79C (the temperature of solid carbon dioxide or "dry ice") for periods up to seven years, and thawed with a high survival rate. But it is interesting to note that a little deterioration occurs even at this temperature; lower temperatures improve the results. (110) It is also observed, contrary to the experience with vinegar eels, that too rapid freezing can be harmful. (110)

Human spermatozoa, without protection, show resistance to extreme cold which varies from cell to cell, and also from donor to donor. In one study, up to 10 per cent of the sperm cells survived five-minute exposure; hardihood varied from donor to donor, but for a single donor survival was the same at -79C, -196C, and -269C. (110)

Dramatic evidence of the viability of deep-frozen human sperm is furnished in a New York Times Service article (Detroit Free Press) of September 6, 1963. Two babies were born to women who had been artificially inseminated with sperm stored for two months at liquid nitrogen temperature. Dr. Jerome K. Sherman, of the University of Arkansas, is said to have stored semen at this temperature for three and a half years without loss.

當我們將注意力轉移到較大型而且較高度進化的生命型態上時，我們也可以看到許多細胞和組織甚至於器官，在經過冰凍之後而復甦的成功案例。然而其中通常都需要用到一些保護製劑，但是並非所有的案例都需要。

公牛的精液在經過甘油的處理過後，曾經被儲存在攝氏零下 79 度 (二氧化碳的凝固溫度或是"乾冰") 長達七年之久，然後在經過解凍之後，其存活率還是非常地高。值得注意地，甚至於在這種溫度下，其間所產生的變質都可以如此地小，何況在更低的溫度下，其結果一定會更加獲得改善。(110) 此實驗也觀察到一個和醋蛆實驗相反的現象，也就是冰凍的速率如果太快可能會是有害的。(110)

人體的精子在沒有施加任何的保護之下，其對極度低溫的抵抗力乃是會因著細胞而異，而且也會因著精子提供者而異。在一個研究中發現，高達百分之十的精子細胞，在經過五分鐘的暴露之後還能夠存活，其堅強度乃是會因供精者而異，但是對單一供精者而言，其堅強度在攝氏零下 79 度，196 度和 269 度都是一樣的。 (110)

人體精子經過深度冰凍後而能存活的顯著證據，在1963年九月六日的紐約時報報系 (底特律自由時報) 的一篇文章中有所報導。一個婦女用被儲存在液態氮中已經超過兩個月的精子來進行人工授精後，竟然還可以生出兩個嬰孩。據報導，阿肯薩斯大學的Jerome K. Sherman博士已經可以將精液儲存在這種溫度下，歷經三年半而無傷大雅。

Dr. S. W. Jacob and co-workers have reported cooling human conjunctival cells (from the membrane lining the eyelid) as well as sperm to within less than one degree of absolute zero, with viability maintained. (50)

Embryo chicken hearts, after treatment with glycerol solution, have been cooled to -190C, and heating resumed after thawing. This was one of the developments which led Dr. D. K. C. MacDonald of Ottawa University, an expert in low-temperature physics, to write, ". . . perhaps the day will come when, if you want it, you can arrange to 'hibernate' for a thousand years or so in liquid air, and then be 'wakened up' again to see how the world has changed in the meantime." (65)

In the case of the mammals, attempts to freeze, store, thaw, and revive specimens have not yet been completely successful. But there have been many partial successes, and much has been learned.

The best-known experiments may be those of Dr. Audrey U. Smith, of the National Institute for Medical Research, Mill Hill, London, working with golden hamsters. These animals have been successfully revived after being about half frozen. In particular, more than half the water in the brain had changed to ice, and the bodies were rigid; yet these mammals recovered to apparently normal activity. (110) This is very important, since it seems to provide some evidence that mental faculties can survive freezing and thawing.

S. W. Jacob 博士和其研究團隊發表，將人體結膜的細胞 (從眼睫內膜取的的) 以及精子冰凍到和絕對零度相差不到一度的範圍內，其活性都可以被維持住。(50)

小雞胚胎的心臟，在經甘油溶液的處理過後，已經可以被冰凍到攝氏零下 190 度，然後經過解凍而恢復其活溫。渥太華大學的 D. K. C. MacDonald 博士乃是一個低溫物理學的專家。鑒於這類的研究進展導致他有感而發，"..........或許這樣的日子將臨到，假使你想要的話，你可以安排自己來在液態空氣中 '冬眠' 大約一前年左右，然後再度被 '喚醒' 來看看到時候的世界到底會變成怎樣。" (65)

有關嘗試來冰凍，儲存，解凍和復甦哺乳類動物樣本，到現在為止還沒有完全成功的案例。但是倒是有許多部分成功的案例，而且從其中我們也學到了許多的知識。

其中最有名的可能就是英國倫敦Mill Hill國家醫學研究中心Audrey U. Smith博士，針對研究金黃色天竺鼠的實驗了。這些動物在經過大約半度的冰凍之後，已經能夠成功地被復甦。猶其是在其腦部中，有超過一半的水分都已經被轉變成冰，而且其軀體都已經凍硬的狀況下，這些哺乳動物都還能夠恢復成看起來具有正常的活性。(110) 這結果是非常重要的，因為這似乎提供了心思功能可以經得起冰凍和解凍過程的某些證據。

It is to be noted that Dr. Smith's results were achieved by crude means: the cooling was with cold baths and cold packs, and the aids to resuscitation were simply artificial respiration and microwave diathermy. The tissues were not given any local protection in the form of special infusions, although it is known that such protection can be very important.

Similar work includes that of Andjus and Lovelock, who have reported recovery and long-term survival of 80 per cent to 100 per cent of ice-cold rats. (110) Dr. J. R. Kenyon and his co-workers have chilled dogs approximately to the freezing point, with heartbeat and circulation completely stopped, and obtained sufficiently complete recovery so that they survived many weeks after the experiment. Chemical infusions were used to counter-act accumulation of certain harmful metabolic products. (55)

The mechanism of freezing damage is still poorly understood. There is much variation in hardihood among different types of cells, and even among individual cells of the same type. Different temperature ranges also have their own distinctive problems.

Experimental work directed toward testing new theories and new protective agents and techniques proceeds vigorously, but on a relatively small scale. When the public becomes interested in freezers, progress should become much swifter. It is not always possible to hasten scientific progress simply y spending more money, but in this instance the possibility seems to exist. Many avenues apparently are not being explored, for lack of workers.

值得我們注意的，Smith 博士乃是用粗糙的方法來獲得其實驗結果的：其冰凍方法乃是用冷浴和冷包，而其復甦方法乃是用簡單的人工呼吸和微波透熱法。其中的組織都沒有給予任何特殊浸漬形式的局部保護，儘管科學家都知道這種保護可能是非常地重要。

類似的研究還包括有 Andjus 和 Lovelock 的，他們都發現了冰凍過的老鼠有百分之 80 到百分之 100 都可以被復甦而且能夠長期地存活。(110) J. R. Kenyon 博士和其研究同僚也曾經將狗冷卻到大約接近冰點，其心跳和血液循環完全被終止，然而都能夠獲得充足完整的復甦，並且能夠讓其在實驗之後還能夠活到好幾個星期。此實驗中是有用到化學浸漬法來對抗某些有害新陳代謝產物的累積。 (55)

有關冰凍傷害的機制，我們其實還是一知半解。在細胞的各種不同型態之間，其堅韌度還是存在有極大的差異，甚至於在同一型態之間，其單獨細胞之間也是一樣。而在不同溫度範圍之間，也都存在著其獨特的一些問題。

專注於試驗新理論，新的保護製劑和方法的研究工作，目前正如火如荼地在進展中，然而其規模都相對地小。如果當社會大眾對冰凍技術都變得非常有興趣時，其進展可能就會變得較快。一般科學的進展是不可能光靠花上大把銀子就可以使其加速進展的，但是就這個例子而言，此可能性倒卻是存在的。有許多途徑很明顯地就是沒有被探索過，只是缺乏對其研究的工作者。

Among other things, a massive, systematic search for new protective agents seems called for.

Even with work at the present relatively slow pace, there is much optimism. Dr. A. S. Parkes, F.R.S., in the foreword to Dr. Smith's hook, says that in the next decade (1961-71), "The preservation [in deep freeze] of whole organs for transplantation may become possible . . ." (110)

Dr. Juan Negrin, Jr. (Lenox Hill Hospital, New York) is reported in 1961 as saying, "We are working now to develop a method for using full body freezing to suspend life. We have already succeeded in bringing about this state in various animals." (117) Some new successes will no doubt be on a cut-and-try, empirical basis. But in order to get a better idea of future prospects and present possibilities, let us briefly review current ideas about freezing damage.

The Mechanism of Freezing Damage

There are several suspected reasons for the frequent failure of animal cells and tissues to survive after being cooled to very low temperatures, stored, and thawed.

Before listing these possible causes of freezing damage, it should be pointed out that "failure to survive" is a very vague and possibly misleading expression.

除此之外，針對一些新的保護製劑，看起來我們似乎還需要有一個大型的系統性的研究。

　　儘管目前的研究步調是相對地緩慢，但是其中卻還是充滿了樂觀。在 Smith 博士書中的前言裡，A. S. Parkes 博士 (F. R. S.) 表示在未來的十年中 (1961–1971)，"整個器官 [在深度冰凍下] 的保存以供器官移植，可能會變成可能.........." (110)

　　紐約 Lenox Hill 醫院的 Juan Negrin, Jr.博士在 1961 年被報導表示，"我們現在正在開發一種方法，要用全身冰凍技術來暫停生命。我們已經成功地在許多不同的動物上能夠達到這種境界。" (117) 其間某些新的成功案例，無疑地將會僅僅是屬於瞎貓抓到死耗子的試驗性本質。但是為了要對未來的前景和當今的可能性能夠有較好的理解，就讓我們一起來檢視當今有關冰凍傷害的一些觀念。

冰凍傷害的機制

　　針對在經過冰凍到非常低的溫度，歷經儲存和解凍過後，動物細胞和組織經常無法生存中，存在著許多臆測的原因。

　　在還沒有去聽這些冰凍傷害的各種可能原因之前，我們應該要指出，"無法生存"乃是一種非常模糊而且有可能是一種誤導的說法。

The usual criterion for survival is resumption of function, if an entire organ is involved, or growth in culture or successful transplant or autoreplant if a piece of tissue is in question. (Autoreplant refers to grafting the tissue back into the donor animal.) A tissue just below the borderline of resumed function is called "dead," and an experiment in which only a small percentage of the cells survive may be considered a failure. But in fact, near successes and partial successes afford substantial grounds for optimism, since they suggest that a comparatively small amount of damage has been done.

It is convenient to list several separate types of possible freezing injury, even though they are not all mutually exclusive, as follows:

1. There may be mechanical damage by ice crystals.

The most obvious opportunity for injury would be a stabbing, crushing, or bursting action against the cell membranes and cell bodies by the ice crystals formed as water freezes. Yet oddly enough, this kind of event seems rarely to have been observed, although it may sometimes occur. (In the case of plant tissues, with their more rigid membranes, this kind of damage occurs much more easily.)

In slow freezing - involving a typical cooling rate of, say, one Centigrade degree per minute - pure ice gradually separates out from the solution in the cell, the ice crystals forming beyond the membrane in the intercellular spaces. Slower freezing produces crystals that are larger in size, and of course fewer in number; faster freezing, the reverse.

通常所謂生存的標準，假使牽扯到的是一整個器官，乃是指其功能的恢復，假使是有關一塊組織的問題，則是指其培養中的成長，或是成功的移植或是自身再植。(自身再植乃是指將組織接植回去原組織供應動物。) 一個組織如果剛好低於功能恢復的界線，就被稱為"死亡"，而在一個試驗中如果僅有一小百分比的細胞得以存活，卻可能僅被認為是一個失敗。但是事實上在接近成功和部份成功之間，還存在著相當大值得樂觀的空間，因為其間意味著相對上有較少量的傷害產生。

雖然下列冰凍傷害的可能並非互相都沒有重疊之處，但是現在還是將其許多不同的型態臚列出來，以方便大家:

1. 可能會有由冰晶造成的機械性傷害

最明顯的傷害機會乃是當水冰凍時所產生的冰晶，對細胞膜和細胞本體所進行的一個刺破，壓破，或是脹破的行為。但是很奇怪地，雖然這種情形有時可能會發生，可是它看起來卻似乎很少被觀察到過。(就植物的組織而言，因為其細胞膜較為剛硬，所以這類型的傷害就更容易發生。)

在緩慢冰凍過程──其間一般的冰凍速率假使是每分鐘降低攝氏一度──在細胞中純冰會逐漸地從溶液中分離出來，冰晶會在細胞膜之外的細胞間空間漸漸形成。越慢的冰凍過程就會產生較大顆的冰晶，而當然其數量就會較少。越快的冰凍過程，則反之。

When the so-called eutectic temperature is reached, the remaining solution freezes out in a close mixture of crystals of ice and of the various salts or their hydrates.

There is ample evidence that ice crystal formation as such is not necessarily fatal, even though water expands when it freezes. Meryman says: "Experimental frostbite research produces evidence that a dog's leg can survive after the deep tissues have been at a temperature well below freezing for as much as fifteen to thirty minutes . . . There is no question but that ice crystals are formed, and yet the tissue survives . . . there appears to be little question but that in the soft tissues encountered in the animal kingdom it is possible for an ice crystal to intrude itself between the cells and to collapse the cells completely without impairing their capacity for survival." (70)

In fast freezing, the crystals formed are much smaller, and possibly for that reason less dangerous mechanically, even though the total volume of ice is the same. But fast freezing does not allow the water to leave the cell, and small intracellular or even intranuclear ice crystals may form, with poorly known but probably dangerous potentialities. For example, a membrane surrounding the cell nucleus may be violated.

2. There may be a dangerous concentration of electrolytes.

Since freezing involves a separation of ice from solution, it is a process of dehydration. The fluid left behind in the cell has an unnaturally high concentration of salts and similar substances, called "electrolytes," which have special electrical and chemical properties. This drastically changed internal environment may be fatal to the cell. (69)

當到達所謂的共熔溫度時，所剩下來的溶液就會以冰和各種不同鹽類或是其水化合物結晶形成的緊密混合物，而被凍析出來。

　　雖然當水冰凍時會膨脹，但是有大量的科學證據證明，像這樣冰晶的形成並不見得一定會是致命的。**Meryman** 表示："實驗性的凍瘡研究發現，在溫度大大小於冰點下，一隻狗腿部的深層組織經過大約 **15** 至 **30** 分鐘後，還可以存活的證據……針對冰晶雖然會產生，然而組織還是可以存活這點是沒有什麼可質疑的……但是針對動物界中所遇到的一些軟性組織，冰晶可能會侵入其細胞之間，而將細胞完全地崩解後，仍然不會損害其存活的能力這點，看起來倒是有一點問題。"**(70)**

　　在快速冰凍時，其所產生的結晶會小許多，也可能是因為這樣，雖冰的總體積是一樣的，然其機械性上的危害還是較少。但是快速冰凍無法讓水分有離開細胞的機會，因此有可能產生出在細胞內，甚至於細胞核內的微小冰晶，這將會有雖然尚未被知悉、然而卻可能頗具危害性的各種潛力。例如，環繞細胞核的薄膜可能就因而被侵犯了。

2. 可能會產生一個危害性的電解質濃度

　　既然冰凍過程中牽扯到冰從溶液中分離的事實，其實這就是一個脫水的過程。殘留在細胞中的液體會具有一個不正常高的鹽分和類似物質，也就是所謂 "電解質" 的濃度，這是具有其特殊的電學和化學性質的。這種內部環境劇烈的轉變，對細胞而言可能是致命性的。**(69)**

Damage to the cell from this cause is thought to be dependent on the degree of electrolyte concentration, the time of exposure to it, and the temperature; a lower temperature means a slower reaction. The electrolyte concentration may be dangerously high, depending on the type of cell and other factors, roughly between 0C and -25C. Hence cooling in this range should be relatively rapid, if possible, in the absence of protective infusions.

Dr. J. E. Lovelock thinks the lipoproteins are especially sensitive to denaturation, or loss of chemical characteristics, from this cause. "A frequent if not invariable component of the many membranes of a complex living cell is the lipid-protein complex . . . held together not by the relatively strong covalent bonds which link the atoms of a simple protein, but by weak association forces similar to those supporting a soap bubble, these complexes are inherently unstable and probably maintained in living cells by continuous synthesis . . . Freezing [can easily] denature the more sensitive lipid-protein complexes of the cell.

"The high sensitivity of lipid-protein complexes to the adverse effects of freezing suggests that not only the principal cell membrane, but also the lesser membranes of the cell . . . may suffer irreversible damage during freezing. The profound change in the environment of the cell which occurs during freezing is also capable of causing harm to the more stable molecular constituents of the cell." (62)

由這種原因所造成的傷害，科學家認為是會依電解質濃化的程度，暴露於其中的時間和溫度而定。溫度較低，意味著其反應會較慢。因著細胞的種類以及其他因素，電解質的濃度大約在攝氏零度到零下 **25** 度時，就可能有高到非常危險的境界。因此在沒有任何保護性的浸漬時，如果可能的話，在此溫度範圍時的冰凍，其進行相對上應該要快一點。

　　J. E. Lovelock 博士認為從這種成因中，脂性蛋白質是特別容易產生變質，或是失去其化學上的特性的。　”在一個複雜的活細胞中，其上許多胞膜的成分，假使不是一成不變地，也經常會是脂肪－蛋白質複合物，⋯⋯不是由一般相對上較強來連結簡單蛋白質的原子共價鍵，而是由類似支撐一個肥皂泡漠的微薄連結力量，來結合在一起的，這類複合物在先天上是較不穩定的，而且在活細胞中可能必需藉由繼續不斷的合成來維持，⋯⋯⋯冰凍 (很容易地就會) 使細胞中較為敏感的脂肪－蛋白質複合物變質。”

　　”脂肪－蛋白質複合物對冰凍效應的惡性高敏感度，意味著不僅是主要的細胞膜，就算是細胞內較次要的各種薄膜，⋯⋯在冰凍過程中都可能會遭遇到回天乏術的傷害。在冰凍過程時，細胞內在環境的深度轉變，對細胞內那些較穩定的分子組成物也有可能會造成傷害。” (62)

Not to put too fearful a face on it, we should note also that he goes on to say, ". . . many living cells and tissues have now been stored successfully in the frozen state ... in spite of these formidable hazards."

We should also remind ourselves once more that the phrase, "irreversible damage" is used much too cavalierly, and really means only "incapable of being reversed by methods so far employed."

3. There may be metabolic imbalance.

Dr. L. R. Rey, a prominent investigator of the Ecole Normale Superieure, Paris, believes the cells may be thrown out of kilter by the unequal effect of cold on delicately balanced life processes. "Various enzymes are not inhibited in the same manner ... there may be an abnormal accumulation of intermediate metabolites which normally have a transitory existence and which may either prove to be toxic or to orient the metabolism in a different direction." (90)

This sounds rather hopeful, since it seems to leave open the possibility of redressing the balance, once we have both the understanding and the means.

為了要避免過度恐懼地描述，我們應該要注意到他也繼續地提到，"………儘管有這些可怕的危害，……現在已經有許多活的細胞和組織都已經被成功地儲存在這種冰凍的狀態。"

我們也應該再度提醒我們自己，所謂 " 回天乏術的傷害" 的詞句乃是被用得有點輕率，而其真正的意涵乃僅是指 "應用目前的方法無法來回復。"

3. 可能會造成新陳代謝不平衡

L. R. Rey 博士乃是法國巴黎 Ecole Normale Superieure 的一個傑出的研究科學家，他相信在非常精細微妙地平衡的生命現象中，細胞可能會因著冰凍的不均衡效應，而被搞得失去正常秩序。"各種酵素不再會以正常的方式受到抑制，……可能會有不正常的新陳代謝中間產物的累積，此中間產物的存在一般都是過渡性的，此物的累積有可能會被證實是有毒的，或是會把新陳代謝導引到另外一個不同的方向。" (90)

這聽起來到還是頗具希望的，因為它看起來似乎還留下一個重新調節平衡的可能性，只要我們能夠對其有足夠的理解並且找到方法的話。

A similar comment has been made by Dr. L. Kreyherg. "It is evident that in areas of organized tissue in situ [on site] the limits for survival of some of the cells after freezing . . . is not decided by the tolerance of the individual cells, but by the local reactions to the disorganization of the social life of the cells." (56) One suspects a like remark might apply to conditions within an individual cell and between its parts.

4. There may be thermal shock and osmotic shock.

Rapid freezing is fatal to many cells, for reasons not well understood. One hypothesis about "thermal shock" is that various materials in the cells and their membranes shrink at different rates as the temperature is lowered, setting up destructive mechanical stresses. "Osmotic shock" refers to the unfavorable effects of sudden changes in solute concentrations in contact with certain membranes.

5. There may be damage during storage.

The cell encounters various vicissitudes as it is cooled, depending on many factors in each of several ranges of temperature; and when it finally arrives at storage temperature, its troubles may not be over. As already pointed out, there is evidence that at all but the very lowest temperatures, near absolute zero, eventually appreciable changes do take place, although they may be very slow.

L. Kreyherg 博士也發表過類似的評論。 ”攸關現場排列整齊的組織研究領域，某些細胞在經過冰凍後存活的限制上，很明顯地……乃不是由單獨細胞的耐力來決定的，而是由這些細胞對其社群生活上的失序所產生的局部反應來決定的。" (56) 我們認為同樣的解說也可能推衍到一個單獨的細胞內，以及其間各種組成物之間的情境。

4. 可能會有熱力休克和滲透壓休克

因著某些尚未了解的原因，快速冰凍對許多細胞是會致命的。有關 ”熱力休克” 的一個假說乃是認為，當溫度被降低時，細胞中的各種不同的物質以及其各種不同的薄膜會以不同的速率收縮，因而產生出一些具破壞性的機械應力。” 滲透壓休克” 乃是指因著某些溶質濃度的突然改變，使其在和某些薄膜接觸時會產生出具傷害性的作用。

5. 儲存過程中可能會造成傷害

細胞在其冰凍的過程中會遭遇到各種不同程度的枯萎，這乃是會因著在每一個不同溫度範圍中所存在的許多因素而定。然而就算其最後已經達到儲存的溫度時，其折磨並未因此而結束。就如同我們已經指出的，除非達到一個最最低的溫度，也就是接近絕對零度時，否則最後總是會有一些可觀的變化產生，雖然其可能是非常地緩慢。

Although Fernandez-Moran has pointed out that free radical activity can occur at -196C, and suggested that perhaps long-term storage ought to be at liquid helium temperatures, nevertheless most writers seem to agree that storage at the temperature of boiling nitrogen is probably safe.

In any case, the word "decay" is probably ill-chosen to describe the deterioration that may take place at low temperatures. It is probably not a case of general rot or putrefaction, or even normal metabolism, proceeding at a reduced rate, but rather a case of a few sensitive processes going essentially to completion, with ensuing stability for an indefinite period. If this is true, cooling with dry ice for long periods may be just about as safe as cooling with liquid helium, except for some initial minor damage. On this, however, I cannot quote authority, and many questions remain unanswered.

6. There may be thawing damage.

There is ample evidence that more damage may occur during thawing than during freezing, especially if thawing is slow and protective infusions are lacking. The mechanisms of damage appear to include migratory recrystallization of ice (small crystals may merge into larger crystals, causing mechanical disruption) and gas bubble formation, as well as others. These effects may occur at temperatures as low as -40C.

For a time, the difficulty of obtaining fast thawing was thought to be extremely serious for any but the smallest specimens, for which heat exchange is not a problem.

然 Fernandez—Moran 已經指出,自由基在攝氏零下 196 度時還會有作用,並且建議或許長期的儲存必須是在液態氦的溫度下,但是大部分的科學家似乎都同意,儲存在氮氣的沸點溫度就可能是安全的了。

不管如何,選擇用 ”腐敗” 這個字眼來描述低溫下所可能發生的變質可能是有點不太恰當。因為其可能不是一個全面性的腐爛或是敗壞的情形,甚或不是一個在減慢速率下一般正常的新陳代謝,而是一種僅讓幾個較敏感的反應進行到大致完全的狀況,因而使其可以穩定到一段不定期長的時間。假使這是真實的話,用乾冰來作長期的冰凍,除了一些先期的次要傷害之外,可能會和使用液態氦來冰凍大致一樣地安全。但是針對這一點,我無法找到權威性的引證,而且其中還存在著許多有待回答的疑問。

6. 可能會有解凍的傷害

有許多證據顯示在解凍的過程中會產生比冰凍的過程中更多的傷害,尤其是當解凍很慢而且沒有任何的保護性浸漬時。這種傷害的機制看起來似乎包括有遷移性冰的再結晶 **(migratory recrystallization of ice)** (微小的冰晶可能會合併成較大的冰晶而造成機械性的破壞),以及氣泡的形成和其他的一些原因。這些效應在溫度低到攝氏零下 40 度時都還有可能會發生。

有一段時間,,因為在一些最小的樣本上沒有熱交換的問題,所以除了這些之外,達到快速解凍的困難曾經被認為是極端嚴重的問題。

It now appears, however, that microwave diathermy and induction methods will allow rapid thawing, at a more or less uniform rate throughout the body, even of large specimens. These methods involve the use of high frequency radio waves, alternating magnetic fields, or alternating electric fields; the former are analogous to an ordinary heating lamp, the latter to so-called electronic ovens. Apparatus has been described y Lovelock. (61) Using this, rabbits can be thawed in just a few seconds. (110)

7. There may be miscellaneous deleterious effects.

Various bits of evidence and speculation point to additional possibilities in the complex question of freezing injury. Drugs and antibiotics, as well as normal body solutes, may become concentrated to lethal levels. At dry ice temperature, if glycerol is used, there may be incomplete freezing, and a slight solubility of salts in glycerol may allow slow damage. At extremely low temperatures, complete removal of water as ice might include water molecules necessary for the structural integrity of proteins. And so it goes; much is known, but much more needs to be learned.

In summary, if we seek the main danger to humans frozen without perfusion by protective chemicals, expert consensus seems to point to denaturation of protein molecules, a consequence of overexposure to concentrated salt solutions, which in turn is a consequence of too-slow freezing. As to the possibility of avoiding this danger by using protective agents, or by increasing the speed of freezing, more will be said later.

但是現在看起來，微波透熱法和感應法 (induction method) 已經可以達到快速解凍，甚至於對一些大型的樣本，已經可以有或多或少全身均勻的速率。這些方法牽涉到高頻率無線電波，交流磁場或是交流電場的利用，前者乃是和一般加熱燈泡類似，而後者就是所謂的電子爐。**Lovelock** 對此設備有所敘述。 **(61)** 使用這個，兔子可以在短短幾秒鐘內被解凍。 **(110)**

7. 可能還會有各類的損害性效應

各種片段的證據和臆測都認為,在冰凍傷害這個複雜的問題上還會有其他的可能性。藥物和抗生素,以及一般體內的溶質,都有可能被濃縮到一些變成會致命的濃度。在乾冰的溫度下,假使是使用甘油,就有可能會有不完全的冰凍,而些微溶解在甘油中的鹽類,就有可能會產生緩慢的傷害。在極度低的溫度下,水分以冰的狀態被完全移除,也可能會將維持蛋白質結構完整所需要的水分子移除掉。這類的理由可以有一大堆,其中有很多是已經知道的,但是還有很多是必須要進一步研究的。

總結言之,在沒有用保護性的化學藥物來浸漬下,假使我們要找出人體冰凍的主要危險,專家的共識都指向蛋白質分子的變質,這乃是由於過度暴露於高濃度鹽類溶液的後果,其次這也是冰凍太緩慢的一個後果。有關如何利用保護性製劑,或是藉由冰凍速率的加快來避免掉這個危險,容我們稍後再加以敘述。

Frostbite

We are now in a position to answer the skeptics who say that, because a frostbitten toe may be incurable today, they doubt it will ever be possible to freeze and revive a complete man; and it may be worthwhile to make the answer explicit.

To begin with, frostbite often is cured, as shown by both clinical and laboratory experience. When we investigate which cases are cured and which are not, we find some neat tie-ins with the earlier discussion of the mechanism of freezing damage.

It has been shown, both in man and other animals, that freezing may actually occur, with formation of ice crystals in the tissues, without any irreversible harm. (110) The damage occurs if the temperature is too low, so that too much ice separates out, producing too high a concentration of solutes in the tissue fluids; or if freezing is too protracted, resulting in exposure of the cells to concentrated solutes for too long a time; or if thawing is too slow, resulting in dangerous high-temperature exposure to somewhat concentrated solutes; or if there has been bending or rubbing of the member while frozen, damaging non-resilient tissues; or if unfrozen but chilled and malfunctioning blood vessels fail to supply the thawed parts.

Medical texts recognize that thawing should be rapid, and rubbing (with snow or anything else) avoided. (12)

凍瘡

我們現在已經有立場來回答懷疑者所提出的問題,也就是說因為當今一根被凍瘡的腳指頭可能都是無法痊癒的,因此他們質疑冰凍和復甦一個完整的人將會是有可能的。而這應該值得我們去將此答案說清楚講明白。

首先,依據臨床和實驗室的經驗所顯示的,凍瘡常常是可痊癒的。當我們對痊癒的和沒有痊癒的案例進行研究時,我們發現其中某些和我們前面所討論的冰凍傷害的機制是息息相關的。

科學研究顯示,不管在人類和其他動物,冰凍可能實質上產生過,其中有冰晶在組織中形成,但卻可以是沒有任何回天乏術的傷害。(110) 假使溫度太低,導致有太多冰被分離出來,而在組織液中產生一個過高溶質的濃度;或是假如冰凍拖得太久,導致細胞暴露在高濃度溶質的時間太長;或是假使解凍的速度太緩慢,導致危險性的高溫暴露在濃度有點高的溶質之中;或是假如在冰凍時,肢體有被彎曲或是摩擦,而傷害到不堅韌的組織;或是假使在解凍後,那些被冰凍和失效的血管無法供血到解凍的部位都有可能產生傷害。

一些醫學的教科書都有認知到,解凍一定要快,而且應該要去避免 (用雪或是其他任何東西) 的擦磨。(12)

In a word, the presently incurable cases of frostbite are simply those cases in which the conditions were unfavorable. In other cases, frostbite can be cured. In fact, human skin has been rapidly frozen to dry ice temperature, and then used in grafts with some success. (110) Rabbit skin has been stored at dry ice temperature for four years without deterioration, after being pre- treated with glycerol. (110)

It is not obvious how a whole man could be rapidly frozen, or treated with glycerol, but these matters will be discussed later. The point here is simply that much is known about frostbite, it is preventable, and it is often curable. In addition, of course, some of the cases now thought incurable will be curable in the future.

The Action of Protective Agents

A brief review of the substances which have been found useful as protective infusions to prevent or reduce freezing damage, and of the theory of their action, shows that a good beginning has been made in the research, and that we are not without resources even now.

An ideal protective agent is one to which the cells are readily permeable, which prevents all kinds of freezing damage but is not itself toxic, and which can be easily removed after thawing. Nothing is known which completely fills this bill for all kinds of tissues. The substances which seem to be most nearly and most widely satisfactory are glycerol and dimethylsulfoxide.

簡而言之，當今凍瘡無法痊癒的案例，純粹是由於那些案例所處的狀況都是不利的。不然的話，凍瘡都是可以被治癒的。事實上，人體的皮膚已經有被快速冰凍到乾冰的溫度，然後被成功地用來移植。 (110) 兔子的皮膚在用甘油預先處理之後，曾經被儲存在乾冰的溫度下，歷經四年而沒有任何變質。 (110)

如何來快速冰凍一整個人，或是用甘油來處理，我們還是不太清楚，但是我們將在後面探討這些事宜。在這裡的重點單純地就是，凍瘡已經被深度理解，它是可以被預防的，而且它常常是可治癒的。另外當然，某些現在被認為是無法治癒的，在未來將會是可治癒的。

保護製劑的作用

在對已經被發現能夠在預防或是減少冰凍傷害的保護浸漬上有用的物質，以及它們作用的原理進行一個簡要的評估之後，發覺在其研究上其實已經有一個很好的開端，而且甚至於在現在我們也不是沒有任何資源。

一種理想上的保護製劑，對它而言細胞應該是很容易就可以滲透的，它應該可以防止各種不同的冰凍傷害，但是本身應該是沒有毒性的，而且在解凍之後，它應該可以很容易就被移除。沒有一種已知東西，在用在各種不同的組織上，是能夠完全符合以上的要求的。看起來似乎最接近而且具最廣泛滿意度的東西乃是甘油和二甲基氧化硫 。

Glycerol, in particular, has been extensively tested. Its use has been markedly successful, although not always completely successful, with a wide variety of organisms and tissues, including mammalian kidneys, bone, lungs, sperm, skin, hearts, ovarian and testicular tissue, and - most important - nervous tissue. (110)

In most cases, glycerol is thought to exert its beneficial action mainly by buffering the solution of electrolytes, that is by somehow preventing or reducing the chemical action of the dissolved substances. This action may be linked to the capacity of glycerol to bind water, and itself to dissolve some of the salts. Glycerol also suppresses the occurrence of a sharp eutectic point in physiological media; if there is no sudden crystallization, the cells may be saved from osmotic shock. (110) Other modes of protection may also occur, and the relative importance of the various modes depends on the nature of the tissue.

Other substances, especially various sugars and alcohols, have been used with varying degrees of success.

Many fascinating experiments have been reported whereby tissues have been induced to tolerate glycerol as a result in adjustments of other components of the solution used for perfusion, such as calcium and potassium; and whereby ingenious methods of removing the glycerol have been devised. It is encouraging to note that in a great many cases where unsolved problems remain, it is the thawing phase and the removal of glycerol which seem to present the difficulty. This suggests that our bodies might be frozen and stored in reasonably good condition, so that future technicians would only have to perfect methods of thawing and removing the protective agents, and would not have to perform excessive wonders in reversing freezing damage.

特別是甘油，已經被廣泛地試用過了。其應用，雖然不見得一直都是完全的成功，但是在一個廣泛種類的生物和組織上，其中包括了哺乳類動物的腎臟，骨骼，肺臟，精子，皮膚，心臟，卵巢和睪丸組織，以及最重要的—神經組織，已經都有顯著的成就了。(110)

在大多數的案例中，科學家認為甘油最主要地乃是藉著緩衝電解質溶液，這就是某些程度上藉著預防或是降低一些溶解物質的化學作用，來展現其有益的作用。這個作用可能和甘油具有綑綁水的能力有關，而且甘油本身可以溶解某一部份的鹽類。在生理媒介物中，甘油也可以壓制一個明確共熔點的發生。假使沒有了這個突然間的結晶化，細胞就有可能免於滲透壓上的休克。(110) 其它的保護模式也有可能發生，而且各種不同模式的相對重要性，乃是依組織的本質而定。

還有其他的物質，尤其是各種不同的糖類和醇類，都有被用過，其中都有達到不同程度的成功。

科學界已經發表許多饒趣的試驗。藉此，有將組織誘導來忍受甘油，而導致其對用來浸漬溶液中的其它成分的適應，例如鈣和鉀；也發明了能夠將甘油移除的一些聰明方法。令人感到欣慰的，雖然在許多的案例中仍然有一些尚待解決的問題，但是看起來較有問題的僅是解凍的階段和甘油的移除。這意味著我們的身體是可以被冰凍，而且是可以以相當好的狀態被儲存起來的，因此未來的技術人員僅要去讓解凍的方法和保護製劑的移除更臻完美就可以了，而不需要去在修復冰凍傷害上進行過多的探索。

The Persistence of Memory after Freezing

Some scientists not so long ago feared that even if we could freeze a body, store it at low temperatures and then restore it to active life, the brain would be wiped clean of memories, resulting in a kind of grown infant or idiot. It is obviously of the utmost importance to assure ourselves that this will not be the case.

Everything hinges on whether memory is dynamic or static. In computing machines, there are two general ways to store information: there are dynamic methods, involving oscillations which will die out if the power is turned off, and there are static methods, such as the use of magnetic tape, in which the information remains even though the machine is not turned on. These two possibilities exist for the brain as well.

As recently as 1960, Professor William Feindel of McGill University wrote: ". . . nerve cells have some of their numerous branches turning back to end on the body of the parent cell, so that they actually receive samplings of their own outgoing messages . . . these self-re-exciting nerve loops may keep up a perpetual circular impulse which is the 'memory' of that particular cell ...." (29) But he also pointed out that memories might be related to physical, chemical, or electrical changes at the hundreds of tiny button-like endings covering each nerve cell in the brain.

More recently, however, Professor I. S. Roy John, director of the University of Rochester Center for Brain Research, has written:

冰凍後記憶的保持

不久之前，有一些科學家曾經擔心說假使就算我們可以將一個人體冰凍，儲存在低溫之中，然後將其恢復成活的生命，其腦部的記憶將會被清理的一乾二淨，因而變成一個嬰孩似的大人或是痴呆。很明顯地，不讓這種情形發生乃是我們應該要確保的一件最重要的事情。

所有的事情乃是繫於到底記憶是動態的還是靜態的。在電算機器中，一般有兩種方法來儲存資訊：一種就是動態類的方法，其中所包含的一些震盪，假使電源被關掉的話，就會完全喪失。另外一種則是靜態類的方法，例如使用磁性記憶體，其中的資訊甚至在機器沒有開啟之下也是存在的。在人腦中，這兩種可能性也是存在的。

最近在 1960 年的時候，McGill 大學的 William Feindel 教授曾經寫到，"……神經細胞上許多的分枝有些是會反轉，而最後終結在其母細胞體的上面，因此它們實質上是可以接收到它們自己所送出訊息的一些樣本……這些自我再度激發的神經迴路可能可以一直維持著一個永久的環狀脈衝，這也就是這個特定細胞的 '記憶'……" (29) 但是他也曾經指出，記憶也可能和人腦中，覆蓋在每一個神經細胞上的好幾百個類似按鈕的終結點的物理，化學和電性的改變有關聯。

但是在更近的時候，Rochester大學腦研究中心主任I. S. Roy John教授曾經寫到：

"Ample evidence exists for a two stage process of memory . . . (1) an early consolidation period approximately 0.5-1.0 hour long, in which reverberatory electrical activity probably maintains a representation of the experience, and (1) a long-lasting stable phase, in which experience is stored as a structural modification of some sort." (51)

In other words, very recent memories are dynamic, and this helps to explain the retrograde amnesias sometimes observed after certain kinds of shock or trauma. But most of the memories, the long-term memories, are static. In fact, they are believed to consist of changes in protein molecules in the brain cells. (46)

Many experimental tests have been made. For example, Dr. Smith reports, "We found, in collaboration with animal psychologists, that rats which had been trained to solve problems of finding food in mazes showed no appreciable loss of memory after cooling to a body temperature just above freezing. Activity of the cerebral cortex, as judged by electroencephalograms, ceases at about + 18C in the rat, so that cerebral activity most have been arrested for 1-2 hr. in all the animals tested. After reanimation they were, nevertheless, capable of acting on previous experience. This result was not consistent with the theory that memory depends upon a continuous passage of nerve impulses through actively metabolizing neurons in the brain." (110)

"有大量的科學證據證明記憶的過程存在有兩個階段.......(1)
其中有一個大約是 0.5 到 1.0 小時長的先期鞏固時段，在此時
段中可能是由反覆的電性活動來維持這個經驗的表徵，接著
(1) 則是一個長久持續的穩定階段，於此階段中這個經驗乃是
以某一種的結構改變來被儲存。" (51)

　　換句話說，非常近期的一些記憶乃是動態的，這也可以用
來解釋在經過某類的休克或是創傷之後，有時可以觀察到一些
退化性失憶症 (retrograde amnesias) 的原因。然而大部分
的記憶，也就是長期的記憶，乃是靜態的。事實上科學家相信，
它們乃是在腦部的細胞中有產生了一些蛋白質分子上的改
變。(46)

　　科學界已經進行過許的的實驗性試驗。例如 Smith 博士
所發表的，"和一些動物心理學家一起合作研究後，我們發現
老鼠經過迷宮訓練，能夠解決食物尋找問題後，將其體溫冰凍
到剛好在冰點之上，在其復甦之後，並沒有顯示出任何可察覺
的記憶喪失。由腦電波圖判讀，老鼠中大腦皮層的活動在大約
攝氏 18 度時就停止，因此在所有試驗過的動物中，大部分大
腦的活動都已經被中只有 1 到 2 個小時了。然而在經過復甦之
後，它們都有辦法依著先前的經驗來活動。理論上，記憶乃是
仰賴著一個透過腦中許多神經元，活潑地代謝所產生的一條神
經脈衝的連續通道而來的。這個實驗結果是和這理論不相契合
的。" (110)

There are two other points of great significance concerning memories: each one seems to be stored in many separate locations in the brain, and therefore may withstand widespread damage; and they consist of chemical coding similar to the traces which record genetic and immunological information, and possibly therefore they may be hardy and resistant to damage.

Professor Hans-Lukas Teuber of the M.I.T. writes, "Experiments employing massive cerebral ablations [removal of parts] or multiple transections [cross sections] of cortex . . . show remarkable resiliency of established 'engrams.' . . . The survival of old established traces following hibernation, general anesthesia, or convulsions suggests a mechanism protected against loss in a manner analogous to immune reactions, i.e., by virtue of multiplication of the trace, relatively small size, and considerable dispersal throughout the cerebrum. . . . [Certain experiments may reveal] that biological trace processes are of essentially the same type, whether we are dealing with genetic processes, embryonic induction, with learning, or immunological.(116)

We shall see the importance of this view when we ask how much freezing damage may be tolerable.

The Extent of Freezing Damage

It must be emphasized that freezing damage, especially to the brain, may not be excessive, even though no mammal has yet made a complete recovery after full-body freezing by the rather crude methods so far employed.

有關記憶還有兩個非常重要的點：每一個記憶似乎都被儲存在腦中許多不同的地點，也因此可以忍受大範圍的傷害；而且它們會有類似紀錄基因和免疫資訊的一些跡象，也可能因此它們可能會是堅韌而且可以抵擋傷害的。

麻省理工學院的 **Hans-Lukas Teuber** 教授寫到，"利用大量大腦摘除 [部分移除] 或是皮層多重橫切 [橫斷面] 法的一些試驗……顯示出既有 '印象' 有極顯著的抗耐力"………在凍眠，一般麻醉，或是痙攣之後老的既有痕跡的留存，意味著有一種類似免疫反應模式的一種機制，來保護其免於喪失，也就是藉著痕跡的複製，雖然其大想相對地微小，卻是相當分散在整個大腦之中………[某些實驗或許有揭露出] 在生物中留跡的過程可以說幾乎是同樣的，不管我們是在討論基因上的過程或是胚胎上的誘導，不管是在學習上面或是在免疫學上面。
(116)

當要去探問可以忍受多少冰凍傷害時，我們將會發覺這個觀點的重要性。

冰凍傷害的程度

我們必須要強調，雖然以目前所用相當粗糙的方法，還沒有哺乳類動物在經過全身的冰凍過後，曾經有過一個完全的復甦，但是冰凍傷害，尤其是對腦部而言，可能不會是太嚴重的。

There are several difficulties in freezing large animals. Perfusion with protective agents is not easy, and fast freezing of deep tissues has been regarded as hopeless. It follows that there may be denaturation of protein molecules in the brain by concentrated salts, and this thought has produced much gloom. In the next section it will be suggested that the major part of the freezing damage can, in fact, be avoided. In this section it will be argued that, even if the freezing injury is as severe as it usually seems to be, reasonable grounds for optimism remain.

First, while it may be hard to conceive of a generalized method for reversing protein denaturation, this is by no means the end of the story. For one thing, such a method may very well be devised, despite our inability to conceive it, by the ingenious men and redoubtable machines of the future. After all, in the last century engineers considered a heavier-than-air flying machine impossible; and before 1926, when Sumner isolated urease, it was not even known for sure that enzymes were proteins. (3) Further, as we shall see, the nature and extent of the denaturation is not uniform, and may in some cases be trivial; and the attack need not necessarily be "generalized."

It must be stressed that even crude freezing frequently fails to kill all cells, and that those "killed" suffer varying degrees of damage; this is true even if we fix our attention on a single type of tissue. Also, the most important parts of the cells may be the hardiest.

在冰凍大型動物上還存在著許多的困難點。用保護製劑來浸漬不是很容易；而深層組織的快速冰凍也被認為是沒有希望。因此會認為在腦部一定會有由被濃縮的鹽類所造成的蛋白質的變質，而由這個意念產生出很大的悲觀。在下一個章節中，我們將解釋為什麼大部分的冰凍傷害事實上是可以避免的，在此章節中，我們要討論的乃是，就算冰凍傷害看起來似乎是如一般人所認為地那麼嚴重，其中還是有一些合理的理由來保持樂觀。

首先，雖然要想出一種通用的方法來逆轉蛋白質的變質可能是非常地困難，但是這故事絕非就壽終正寢了。因為有一點，雖然我們目前沒辦法想出來，但是這種方法可能被一些聰明人士和未來的一些偉大的機器來將之設計出來。畢竟在上個世紀，工程師都認為一架比空氣重的飛行機器是不可能的；而且在 1926 年之前，當 Sumner 分離出尿素酵素時，我們都還不知道酵素就是蛋白質。(3) 還有就如我們將會看到的，變質的本質和程度都是不均勻的，而且在某些情形下還是微不足道的，況且其所受的攻擊並不需要被 "以偏概全".

我們必須要強調，就算是粗糙的冰凍也常常無法殺死所有的細胞，而且在那些 "被殺死" 的中間，所受到的傷害程度也會不同。就算我們將注意力專注於單一型態的組織，其結果也是一樣。況且，細胞中最重要的一些部分有可能是最難傷害的。

That some cells survive freezing even when most "die" we note, e.g., from the work of Rey, who rapidly cooled embryo chicken heart tissue: " . . . there is no growth in the cultures without glycerol except sometimes in two or three migrating cells . . . some peculiar cells do survive after the exposure to liquid nitrogen. . . . Why is the main part of the tissue killed by rapid cooling in liquid nitrogen? . . . we think [these alterations] occur during the thawing process." (90)

Even though chickens are not people and hearts are not brains, it is important that some cells survive; we can logically conclude that probably many others almost survived, and could have been rescued by future scientists either before or after thawing by improved methods.

By way of analogy, imagine viewing (from the air) a strafing attack on a column of troops. If none gets up afterwards, perhaps they are all dead. But if even one or two get up, it is highly probable that many others are merely wounded and not killed.

Again, Kreyberg says: "It is evident that through severe exposures to cold, many cells, sometimes most of the cells, succumb. Sometimes single cells, sometimes smaller groups of cells survive and are able to repopulate cultures and even form rather complex transplants, as demonstrated through the experiments with ovarian tissues." (56)

我們發現雖然當大部分的細胞都 "死亡" 了，但是還是有一些會存活下來，例如依據 Rey 所作的研究，他曾經將雞的心臟組織快速地冰凍，".......在沒有用甘油下，除了有時會有兩三個遷移的細胞之外，組織培養中沒有任何的成長..........在暴露過液態氮之後，某些特異的細胞確實可以存活......為什麼主要部分的組織會被液態氮的快速冰凍殺死呢?.......我們認為[這些變化] 乃是在解凍的過程中發生的。" (90)

雖然雞不是人，而且心臟也不是大腦，但是最重要的是有些細胞是可以活下來的。我們因而可以邏輯地推斷有許多其他的細胞可能幾乎是可以活下來的，而且不管是在解凍前或是解凍後，藉著改良過的技術，是可以被未來的科學家救活的。

以比喻的方式而言，想像說 (從空中) 觀看對伊大群軍隊進行砲彈攻擊。之後如果沒有任何一個人站起來，可能表示他們全死了。但是如果有一兩個人站起來了，這極有可能其中有很多其他的人僅僅是受傷而沒有被殺死。

再度地，Kreyberg 說: "經過嚴酷地暴露在低溫後，很明顯地有很多細胞，某些時候大部分的細胞都死亡了。但是如同透過用卵巢組織的一些實驗所顯示的，有些時候有一些單獨細胞，有些時候有一些較小的細胞群卻可以存活，而且可以再度繁殖增生，甚或可以形成一些相當複雜的移植體。" (56)

There is somewhat similar experience with mammalian nervous tissue, which is the most vital concern. Pascoe, working with rat ganglia, found that although one experiment was mainly negative, "One preparation [without glycerol] was stored overnight at - 150C and on warming the post-ganglionic nerve gave a small action potential when it was stimulated directly." (86)

Not only does experiment indicate that some cells survive unfavorable freezing methods, but theory also. The act of freezing will catch various cells in many different environmental situations and at different phases of the metabolic cycle. Some of these are almost sure to be lucky ones.

Further evidence that freezing damage to the brain may be only moderate, even in the absence of protective infusions, seems to be provided by the work of Dr. H. L. Rosomoff, of The Neurological Institute, New York. He produced lesions in dogs' brains by contact of the dura mater (brain integument) with a brass tube containing liquid nitrogen for eight minutes. If the dogs were kept at normal temperature afterwards, they invariably died, and microscopic examination showed "wide-spread destruction of cellular elements, especially the neurons, complete loss of cytoarchitectural markings, . . ." But of seven dogs kept at 25C or less (after the lesions were produced) for eighteen hours before rewarming, two survived, and the others lived five times longer than those not kept hypothermic; furthermore, examination of the lesions showed that: "The cortical architecture was better preserved, cellular elements showed less evidence of injury, albeit definite degenerative changes were found which may or may not have been reversible in nature." (93)

在哺乳類動物的神經組織上也有或多或少類似的經驗，這是最值得深度關切的。研究老鼠神經節的 **Pascoe** 發現，雖然在一次幾乎完全失敗的試驗中，"一個樣本 [沒有用甘油處理] 經過整個晚上被儲存在攝氏零下 **150** 度，再重溫其結後神經節神經時，當其被直接刺激之時，可以發出一股微小的活動電位。" **(86)**

不僅僅在實驗上指出某些細胞是可以經過一些不利的冰凍方法而存活過來，然而在理論上已是如此。在冰凍的作用中將會碰到處於許多不同環境狀況的，以及處於各種不同新陳代謝循環階段的各種細胞。其中某些幾乎一定會是幸運者。

紐約神經學研究中心 **H. L. Rosomoff** 博士的研究似乎提供了證據，近一步證明了腦部的冰凍傷害，甚至於在沒有保護浸漬之下，可能僅是溫和的。他用裝有液態氮的銅管接觸硬腦膜 (腦的體被) 八分鐘，來在狗的腦部製造出創傷。之後，如果將這些狗置放在一般溫度中，它們一成不變地都會死掉，而顯微鏡的簡是都可以看到 "大範圍細胞元件的破壞，尤其是神經元，細胞建構上的表徵完全喪失，......."，但是其中有七隻狗 (在創傷被製造出來之後)，再重溫之前，被置放在攝氏 **25** 度或更低的溫度下 **18** 個小時，其中有兩隻活過來了，而其它的比那些沒有被放置在低溫下的，要多活超過五倍長的時間。還有，對創傷的簡是發現: "腦部皮層的建構被較好地保持，細胞元件顯示有較少的損傷，儘管還是可以發現一些清楚的退化性改變，而其本質上也許可能或不可能被逆轉。" **(93)**

This experiment was not intended to study freezing damage as such, but was meant to investigate the benefit of hypothermia (reduced temperature) in the aftercare of any kind of brain lesion. Nevertheless, the damage to the cells in the lesion region presumably was produced by freezing. This seems to indicate rather clearly that the most serious damage following such freezing may be the result of anatomical and physiological events during and after thawing, and that while frozen the cells were in relatively good shape. As already pointed out, this is very important, since our task need only be to preserve the bodies with as little damage as may be; if necessary, we can leave to the future the problem of proper treatment during and after thawing.

Again, in the case of nervous tissue pre-treated with glycerol, there is evidence that the major difficulty may lie not in freezing and storage, but in the removal of glycerol. Dr. Smith, commenting on the work of Pascoe, who studied rat nervous tissue after perfusing whole rats with glycerol solution, says, ". . . damage to nervous tissue might not be a limiting factor in attempts to resuscitate a whole animal which had been perfused with glycerol and cooled to and thawed from a very low temperature. " (110)

Having gone to considerable pains to show that even crude freezing methods may not kill all cells, and that even many of the "nonsurvivors" may be only slightly damaged, we are now ready to make our conclusions more explicit.

It will be helpful if the reader will tentatively accept two propositions which will be given support in later chapters:

這個試驗本來並不是要研究這方面的冰凍傷害，而是要去探討低溫 (降低溫度) 對任何種類腦部創傷事後治療的效益。不管如何，在受創區域細胞的傷害可以假設是由冰凍所造成的。這看起來似乎可以非常清楚地指出，經過這種冰凍之後最為嚴重的傷害，可能是由解凍中或是解凍後的解剖學上和生理學上的事件所導致的。而當在冰凍中，這些細胞應該是處於相對好的狀態。就如同我們已經指出的，既然我們的任務僅僅是去在可能越少傷害越好之下來保存身體，因此這發現是非常重要的，我們可以將解凍中和解凍後的恰當處理問題留給未來。

　　再度地，有關用甘油預先處理神經組織的方面，已經有證據顯示最大的困難不在於冰凍和儲存，而在於甘油的移除。Smith 博士曾經研究過整隻老鼠用甘油溶液浸漬過後的老鼠神經組織，他對 Pascoe 的研究工作有所評論，他表示，"………神經組織的傷害，在試圖復甦一整隻經過甘油浸漬，而且被冰凍到一個很低的溫度，然後從中解凍的動物而言，不見得就是一個關鍵因素。" (110)

我們已經費了一大把勁來證明就算粗糙的冰凍方法也可能不致於將全部細胞殺死，而且甚至於那些 "無法存活者" 也可能僅是些微地受到損傷而已，現在我們已經可以來將我們的結論更明顯地表示出來了。

假使讀者可以暫時接受兩個前提，這將會有所助益，當然我們會在後面的章節提出其佐證：

(1) Mastery is beginning to be obtained, and will eventually be thorough, over the growth, development, and differentiation or specialization of both genetic and somatic (body) cells. It will become possible to grow replacement parts, large or small, in culture, or alternatively to make the body repair itself by regenerating missing parts. (In the case of the brain, of course, there cannot be complete replacement or regeneration, since this would be equivalent to growing a new individual.)

(2) Wealth and resources will grow in the future at an ever increasing rate, qualitatively as well as quantitatively. In particular, there will be available fabulous machines, capable not only of action on a titanic scale but also of "thought" on extremely high levels and manipulation on microscopic levels. Now, we recall that memories are stored probably as changes in protein molecules in the brain cells, with multiple locations for each trace in many regions of the brain. (And since the memory recordings are thought to be chemically similar to the codings of genetic information, and since the latter is known to withstand liquid helium temperatures, it may be that memories are equally hardy, but we are not depending on this.) Other elements of personality may be represented in a similar way, or they may inhere in larger-scale circuitry, as in the fiber connections among the nerve cells.

There seems a good chance that the supra-molecular circuitry can be read well enough after freezing. Hence it may well be that only a small percentage of the brain cells need escape with little damage; this may be enough for reasonably faithful reconstruction of the brain with freshly generated tissue.

(1) 不管是在基因上或是體細胞上的培養，成長和分化或是特異化上，精湛的技術正在開始被尋獲，而且最後將會臻於完美。未來在培養器中培養出大小不同的替代的部件，或是另一方法，藉著喪失部分的再生來使身體修復自身，將會變成可能。 (就大腦的案例而言，完整的替代或是在生當然是不可能的，因為這會等於是培養出一個新的個人。)

(2) 在未來財富和資源，不管是在質上或是在量上，都將會以越來越快的速度增長。特別是，未來將會有一些美好的機器，它們不僅可以以巨大的規模作工，有些也可以在極度高階程度上 "思考" 和在顯微的尺度中操控。現在讓我們溫習一下，記憶可能是以腦中細胞蛋白質分子的變化來被儲存的，而且每一片段都存在在腦中許多區域中的多重位址上。(而且既然記憶的紀錄在化學上是被認為是和基因資訊的編碼類似，而且既然大家都知道後者是可以抵擋液態氦的溫度的，因此記憶可能也是一樣地堅韌，然而我們並不需要倚靠這個能力。) 人格中的其它要素可能也是以同一方式來被呈現，或是它們可能是藏在於較大規模的電路當中，如同在諸神經細胞間的纖維接點。

在冰凍過後，超級分子電路似乎還極有可能能夠被好好地讀取。因此，搞不好僅要一小部份的腦部細胞能夠受點傷害而逃過一命就可以了；這樣利用新產生的組織，可能就足夠相當可靠地來重建腦部。

The robot surgeons of the future will have powers now only faintly foreshadowed, but beginnings have already been made in cell surgery. Individual cells have been successfully operated upon, e.g., transplanting nuclei into enucleated amoebae, even cross-species! (27) Thus, if brute-force methods are necessary, it is not inconceivable that huge surgeon-machines, working twenty-four hours a day for decades or even centuries, will tenderly restore the frozen brains, cell by cell, or even molecule by molecule in critical areas.

We hasten to add that in all likelihood the methods used will be much more elegant and yet unforeseen. The great chemist, Linus Pauling, speaking in a general sense not so long ago, said, "The great discoveries of the future - those that will make the world different from the present world - are the discoveries that no one has yet thought about. . . . I know . . . that . . . discoveries will be made that I have not the imagination to describe - and I am awaiting them full of curiosity and enthusiasm." (88)

We must also bear prominently in mind that only those frozen in the very near future may be severely damaged; there will soon be accelerated research, and before many years non-damaging techniques should be available. Indeed, a man can probably be frozen right now with comparatively little injury, as we shall see in the next section.

未來的機械人開刀手術的威力，目前雖僅能模糊地瞥其一影而已，但是在細胞手術上已經有一些開始了。一些單一的細胞已經有被成功地手術過了，例如，將細胞核移植到被去細胞核的阿米巴菌中，甚至於有跨種移植成功的！(27) 因此，假使必須要借重動物力量方法的話，利用一些大型的手術機器，每天二十四小時工作，經過幾十年甚或幾百年，在一些嚴重的地方逐一細胞地，甚或逐一分子地來將被冰凍的大腦細緻地恢復，這不是不能想像得到的。

　　我們急著要補說一句，未來所使用的一些方法，將極有可能比所敘述的要優雅許多，而且是前所未見的。偉大的化學家 Linus Pauling，不久之前以淺顯的方式說，"未來的一些偉大的發現，也就是那些將會使世界變成和現在的世界不同的發現，乃是一些現在連一個人都沒有去想到過的發現……我知道……那些……將被找到的發現乃是我沒有任何想像力可以形容的，而我則是充滿了好奇和熱情等待著它們。" (88)

我們也必須清楚的記在心裡，僅有那些在最近的將來被冰凍的，才有可能被嚴重地受到傷害；不久研究將會加快腳步，而在不經過好幾年之前，不具傷害的技術應該就會問世了。說真的就算現在，一個人也可能在相對少的損害下被冰凍，我們將在下一章節中可以知道。

Rapid Freezing and Perfusion Possibilities

Is a high freezing rate, with cooling of many degrees per minute, really out of the question for an animal as massive as man? And what are the chances of giving a complete large organism the protection of perfusion with a protective agent like glycerol?

It seems that in the absence of a protective agent, the brain (and body) should be frozen quickly. This will not prevent all damage, but might reduce the major risk of protein denaturation. How may quick freezing be accomplished?

Merely immersing the head or body, or even the naked brain in a cold bath such as liquid nitrogen will not do it, except for the outer layers. And while methods of heat transfer other than simple conduction do exist, they do not now seem applicable to cooling the body. The only means that seem presently feasible require more brain surface in contact with the refrigerant.

The most obvious method would be to circulate cold fluids through the brain's blood vessels. This in fact is done in open-heart surgery, but at temperatures above freezing. Whether anything could be done in the subzero range is, so far as I know, one of the open questions requiring investigation. It would certainly be difficult, with vessels tending to be brittle and clogged as well as constricted, but it is not obviously impossible.

急速冰凍和浸漬的可能性

以每分鐘冷卻好幾度的高冰凍速率,對一個像人一樣大的動物進行冰凍,難道真的是不可能嗎? 而用類似甘油的保護製劑,來對一個大型動物進行一個完全的浸漬保護,其成功率到底有多少?

看起來似乎在沒有保護製劑下,腦部 (和身體) 應該可以被快速地冰凍。這將不能免於所有的傷害,但是卻有可能降低蛋白質變質的重大風險。現在到底有少次快速冰凍被完成過多呢?

光是將頭或是身體,或甚至暴露的腦部浸泡在如液態氮的冷浴中,除了其較外面的幾層外,是沒有用的。而且雖然在簡單的傳導之外,一些熱交換的方法是存在的,但是現在看起來似乎對冷卻身體是不適用的。當前唯一看起來似乎可行的方法,乃是需要有更多的腦部表面來和冰凍劑接觸。

最明顯的方法乃是要去將一些冷卻液體循環通過腦部的血管。事實上這已經在開心手術中完成了,但是只是在冰點之上的溫度中進行的。至於這是否可以在低於零度之下完成,就我所知,這乃是一個需要進一步探討的開放問題。這當然會是困難的,因為血管會傾向於脆硬,而且會堵塞,同時也會收縮,但是並非明顯地不可能。

Certain heroic measures also suggest themselves. For example, the brain might be "teased" apart into smaller segments which could be cooled more quickly. Or hollow needles carrying refrigerant might be inserted, as into a pincushion; care would be taken to penetrate different regions in the two hemispheres, to avoid destroying the homologous tissues on each side. Or the brain, after cooling, might even be sliced into sections for quick freezing, on the theory that this mechanical damage, although massive by present criteria, might yet be small compared with the damage done by slow freezing, and more easily repaired.

But the method of choice at the present time would seem to be moderately slow freezing after perfusion with glycerol solution.

Apparently there have been relatively few attempts at full-body perfusion. Dr. Smith says, "So far no technique has been evolved for perfusing individual organs or the whole mammal with glycerol and removing it without damage. If this could be done it might be possible to cool the intact mammal to and resuscitate it from temperatures as low as -70C. Long-term storage of frozen mammals might then be considered. It must be emphasized that there is no prospect of accomplishing this in the near future." (110)

The great thing, however, is that we do not need the fullness of this accomplishment in the near future! Whole rats have been perfused, as previously noted, and probably men could be also. The problem of removing the glycerol without damage can be left to the more distant future, along with the problem of repair of those parts not reached or incompletely protected by the glycerol.

某些勇氣可加的方法也自動地浮現檯面。譬如說，腦部可能可以被“拆解”成較小的塊片，這樣可能可以被冷卻的較快。或是可能可以插入攜帶冰凍劑的空心針，就好像在插入一個針蒜一樣；在兩個腦半球中穿刺不同區域時，一定要特別小心，用來避免破壞到在每一邊的一些同源組織。或是在冷卻之後，甚至可以將腦部切成片段以利快速冰凍。在理論上，這種機械傷害，以當前的標準而言雖然是巨大的，但是比起慢慢冰凍所造成的傷害來講可能是微小的，而且較容易來修復。

　　但是在用甘油溶液浸漬之後，目前所選用的看起來似乎是相對上較慢的冰凍方法。

　　很顯然的，在全身的浸漬上，僅有相對上較少的嚐試。Smith博士表示，"到目前為止還沒有已經進化完整的技術，能夠來在沒有傷害之下，用甘油將單一的器官或是整隻哺乳類動物浸漬，並且將之移除。如果這得以成功的話，將完整的哺乳動物冰凍到攝氏零下70度，並且從其中將之復甦，將可能變成可行的。於是可能就可以來考量冰凍哺乳動物的長期儲存。我們必須強調的，在最近的未來並沒有達成的預期。" (110)

　　但是，最好的事乃是在最近的未來我們並不需要去完全地達成這個! 就如同先前所提出的，整隻老鼠已經被浸漬成功，因此搞不好人也可以如此。至於如何沒有傷害地移除甘油的問題，以及修復那些沒有被甘油處理到或是不完全保護到的部分的問題，我們可以將諸留給較長久的未來。

The people who are dying right now cannot, and need not, wait for 100 per cent mastery of the problem.

The Limits of Delay in Treatment

If you have a dying relative, you can probably give him his best chance by obtaining skilled medical help, planned in advance, to prepare, perfuse, and freeze the body. If this kind of help is not available, and you nevertheless want to give him some chance, more desperate measures are required. In any case, it is important to know how soon after death treatment must be started, and this question will now be considered.

Many laymen, and even many physicians, have the impression that the body must be frozen within a few minutes after clinical death in order to have a chance of revival. This is an error.

It is quite true that if the oxygen supply is cut off, the brain ordinarily seems to suffer damage within three to eight minutes. But this seemingly simple statement is very deceptive: the words "ordinarily" and "damage" both require clarification.

If death comes unexpectedly or without preparation, the brain may certainly suffer "irreversible" damage. When the blood circulation stops, there is no more delivery of oxygen and dextrose, and no more removal of waste products. The immediate causes of damage, according to Wolfe, include increase in inter- and/or intracellular fluid, loss of tone in the capillaries, increased permeability of tissues lining the blood vessels, disturbance of fluid balance, and concentration of lactic acid. (129)

那些現在正在面臨死亡的人，不可能也不需要，去等到這個問題能夠被百分之百的掌握。

處理延遲的限制

假使你有一個瀕臨死亡的親戚，你可能可以由獲得高明的醫療幫助，並且預先規劃其身體的處理，浸漬和冰凍，來給他最好的機會。假使這種幫助無法獲得的話，但是無論如活你還是想給他某些機會，那就需要靠一些較為急切的手段了。就任何一個案例而言，知道死後多快就必須要啟動處理是非常重要的，而讓我們現在就來考量這個問題。

有許多外行人士，甚或有許多醫生都認為，為了要有一個再活的機會，身體在醫學上死亡之後幾分鐘之內就必須要被冰凍。這是一個錯誤。

假使氧氣的供應被切斷的話，通常腦部似乎就會在三分鐘到八分鐘之內蒙受傷害，這乃是不爭的事實。但是這個看起來似乎單純的敘述卻是非常地誤導："通常"和 "傷害" 這兩個字眼都需要澄清。

假使死亡是突如其來或是毫無戒備下的，腦不可能就醫定會蒙受 "不可逆的" 傷害。當血液循環停止時，就不再有氧氣和葡萄糖的輸送，而且也不再有廢物的移除。依據Wolfe，這些傷害的直接成因包括有細胞內和細胞外液體的增加，毛細管內特質的喪失，血管內襯組織滲透性的增高，體液平衡的擾動，以及乳酸濃度的增高。(129)

How quickly the damage begins to occur is not entirely clear. Total circulatory interruption is considered dangerous after three minutes, and the most commonly mentioned limit of tolerance of the brain to lack of oxygen is perhaps five minutes. But Brockman and Jude have conducted experiments with dogs indicating that ten minutes of oxygen deprivation causes no harmful effects, although fourteen minutes at normal body temperature is fatal. They believe the shorter estimates result from use of methods which leave circulation depressed after the period of anoxia, producing added damage and causing the experiments to be misinterpreted. (10)

Of course much depends on temperature and individual variation, as well as other factors. In a later chapter we shall recount the story of a boy who made a nearly complete recovery after twenty-two minutes under water and 2 ½ hours of clinical death.

While the brain cells do indeed "die" more quickly than cells of other kinds, we must not therefore come to a hasty pessimistic conclusion. As already indicated, it may well be that the most important parts and functions of these cells are not so delicate as the cell as a whole.

By way of crude analogy -- which of course must not be stretched too far -- consider a bicycle and a huge snowball rolling down a slope. The bicycle is much more complex, and it can be stopped by merely thrusting a stick through the spokes, while much more effort is needed to stop the snowball. Just the same, a bicycle is on the whole much sturdier than a snowball, and when the stick is removed it will be ready to roll again.

至於傷害多快就會開始發生，則還是不完全清楚。整個循環系統的中斷，經過三分鐘之後就被認為是危險的，而最常被提到的腦部缺氧忍受極限，可能是五分鐘。但是 Brockman 和 Jude 對狗所進行的實驗顯示，雖然在正常體溫下缺氧十四分鐘是會致命的，但是缺氧十分鐘並沒有造成傷害性的效應。他們相信這些較短的估算乃是由於在缺氧時段之後，所使用的一些方法會讓循環受到壓制所導致的，這會產生附加的傷害，而造成這些實驗被錯解。(10)

　　當然這會和溫度，個體間的差異，以及其它的因素相當有關。在後面的章節中，我們將會再度敘述一個男孩的故事，他曾經經歷過22分鐘的淹沒，以及兩個半小時的醫學臨床死亡，而竟然幾乎能夠獲得完全的復活。

雖然腦部細胞的確會比其它種類的細胞 "死亡" 得比較快，但是我們不應該因此就遽下一個悲觀的結論。如同我們已經指出的，這些細胞中最重要的一些部件和功能，大可能不會如細胞的一個整體那麼地脆弱。

　　藉著粗糙的類比，這當然不應該被過度地延伸，試著思考一台腳踏車和一個巨大的雪球同時滾下一段斜坡。雖然腳踏車的結構要複雜許多，但是它僅要穿過輪幅插入一根棍子，就可以被停住，然而要能夠停住那個雪球，就需要耗費更多的功夫了。同樣地，一台腳踏車整體而言是比一個雪球要堅固許多，因此只要將棍子移除之後，它馬上就可以再度運轉了。

It is possible, then, that hope should not be given up so long as any of the body cells show life. If the skin, for example, is still alive, then there is some chance that the brain cells are also alive, albeit damaged. Removal of excess lactic acid, adjustment of the fluid balance, and so on, by techniques at the disposal of future science, may find them good as new.

The period of grace before all of the body cells die is measured at least in hours, and perhaps in days. According to Lillehei et al., the stomach remains alive and healthy outside the body, even without cooling, for at least two hours. (59) Gresham, referring to an unpublished study by V. P. Perry, says, "Tissues removed from cadavers as late as 48 hours postmortem have, in most instances, shown cellular outgrowth in tissue culture. Although this does not eliminate the possibility of cellular alteration, it suggests that many tissues may remain functional for relatively long periods after death, and that postmortem tissues may be satisfactory for viable grafting." (36)

Boiling all this down to a rule of thumb, what perhaps emerges is that in a self-reliance situation, if you want to give the deceased the benefit of even a relatively slim chance, a body should be frozen if it is found on the day of death. If the body has been exposed to cold weather, perhaps the chances are not too remote after two days. It seems entirely possible that the delay damage will still be no greater than the damage of the crude freezing method you may have to use.

因此只要任何一些身體細胞還有生命跡象的話,我們可能就不應該放棄希望。例如,假使皮膚還能活著的話,那麼腦部細胞雖然受損了,但是卻也有還是活著的機會。藉著未來科學界能夠掌握的科技,來將過多的乳酸移除,調節體液的平衡等等,可能可以讓它們和新的細胞一樣好。

在所有身體細胞死亡之前的容許時段至少有好幾個小時,甚或可能有好幾天。依據Lillehei等人,胃部在體外,甚至於在沒有冰凍下,至少可以維持兩個小時的活性和健全。(59)參考 V. P. Perry 沒有發表的研究,Gresham 表示,"從死後已經有48小時的屍體中取出的組織,在大部分的例子中,都顯示出在組織體中都還有細胞上的增長。雖然這沒有排除細胞上轉變的可能性,但是這也意味著在死亡之後,許多組織可能可以維持其功能一段相對長的時間,而這些死亡者的組織可能還足夠用在有效的移植上。" (36)

將所有這些縮減成一個常規,其可能浮現的就是在一個自我倚靠的情形中,假使你要給亡者一個甚至於是相對微薄機會的益處的話,就應該在其發現死亡的當天將其身體冰凍。假使身體曾經暴露在寒冷的天氣,或許在經過兩天之後其機會還不是太渺茫。延遲上的傷害看起來似乎完全有可能不會比你必須要使用的粗糙冰凍方法所造成的傷害要來得嚴重。

In a hospital situation, with medical cooperation, the story is different and much more hopeful, and further remarks are called for.

The Limits of Delay in Cooling and Freezing

Three separate phases of postmortem care of the body may be distinguished: measures in advance of cooling, cooling down to the freezing range, and cooling down to storage temperature.

For various reasons, death may come before cooling equipment is ready. Looking for means to prevent deterioration meanwhile, we find interesting possibilities. Some of them are relevant only if specialized equipment and personnel are at hand, while other measures can be employed by almost anyone.

Methods are already being applied to keep freshly dead bodies in good condition, for the purpose of maintaining organs in good health, when a transplant is contemplated but cannot be performed immediately. Heartlung machines have been used to keep the body supplied with oxygenated blood for up to eighteen hours after death, and then livers taken from the bodies and used for grafts. (Detroit Free Press, October 31, 1963.)

An obvious resort in emergency is to use artificial respiration and external heart massage. (At the same time, the body could be cooled with ice packs, or by exposure to cold air.) Anyone can learn the techniques, and tubes are available so that mouth-to-mouth artificial respiration can be given without actual contact.

在醫院中的情形，因為有醫療的協助，整個故事就不同了，而且是有更大的希望，在此呼籲對其有更進一步的探討。

冷卻和冰凍延遲的限度

身體死後的處置可能可以被細分成三個不同的階段：冷卻之前的措施，冷卻到冰點的範圍，和冷卻到儲存的溫度。

因為各種不同的理由造成在冰凍器材還沒有準備好之前，死亡可能就已經來到。在尋找預防壞掉的方法之時，我們發現了一些有趣的可能性。它們其中有的僅是假使特殊器材和人員齊備時才有搭嘎，而另外的措施則幾乎是可以被任何一個人所應用的。

當準備要進行一項器官移植但是又無法馬上進行時，為了要維持器官能夠處在健康狀態，已經有一些方法被用來保持剛剛死掉的人體在好的狀況。例如心肺機器已經被利用來在死後維持身體充氧血液的供應高達十八個小時以上，再從此身體中摘取肝臟，然後用來移植。(底特律自由日報，一九六三年十月三十一日。)

在緊急狀況下的一個明顯的處置方法乃是去利用人工呼吸和心臟外部按摩。(同時身體可以用冰袋，或是暴露於冷氣中來冷卻。) 任何人都可以學會這些技術，而且目前也有一些管子，所以使得在沒有實際的接觸下，還是可以施行口對口的人工呼吸。

Effectiveness would depend strongly on the cause of death and the condition of the body, but in some cases these simple measures might keep a body in reasonably good condition for hours. In other cases supporting measures might be needed, possibly including injection of anticoagulants.

In certain types of chest injury, possibly help might be obtained from a technique developed by Neely and coworkers. They perfused dogs with a buffered glucose solution instead of blood, and found that ". . . animals can survive 30 min of asanguineous perfusion with no oxygen, and that the survivors exhibited no gross brain damage." (80)

Another intriguing possibility, if the equipment were available, is suggested by the work of Dr. I. Boerema of the University of Amsterdam, The Netherlands. He has obtained some remarkable results treating patients inside a pressure caisson; the surgeons and attendants breathe air at three atmospheres pressure, while the patient breathes pure oxygen at the same pressure. It has been found that the blood circulation can be arrested without harm for about twice as long as normal; at 14.5 deg C dogs can be kept a half hour or longer without extracorporeal circulation. Animals can actually live without blood; with pigs, the hemoglobin could be reduced virtually to zero for at least fifteen minutes, dissolved oxygen taking the place of oxygen carried by red corpuscles.

其效果乃是會和死亡的導因以及身體的狀況有強烈的關係，但是在某些案例中，這些簡單的方法可能可以將一個身體維持在相當良好的狀態下好幾個小時。在其它的案例中，可能還需要一些輔助的措施，可能還包括有抗凝結劑的注射。

　　在某些類型的胸部損傷中，可能可以從一種由 Neely 和其同僚所開發出來的技術中獲得幫助。他們用一種緩衝過的葡萄糖溶液，而不用血液來對狗浸漬，而發現 "……在沒有氧氣之下，動物可以在不同類型物質的浸漬中活過30分鐘，而且在存活者中並沒有顯示出嚴重的腦部傷害。" (80)

　　假使器材齊備的話，荷蘭阿姆斯特丹大學的 I. Boerema 博士的研究中有提出另外一種引人入勝的可能性。他從在壓力沉箱內治療病人中獲得了一些很不錯的結果；外科醫生和其助手呼吸的是三個大氣壓的空氣，而病人所呼吸的乃是同壓力的純氧。科學發現血液循環在不產生傷害下，可以被中斷比正常長兩倍的時間；在攝氏14.5度，沒有體外循環之下，狗可以支撐半個小時。有一些動物實際上是可以沒有血而活著；以豬而言，其紅血球幾乎是可以減少到零至少十五分鐘，溶解氧氣會取代紅血球所攜帶的氧氣。

"... when an animal or patient breathes pure oxygen at 3 atmospheres (absolute) there is a greatly increased physical solution of oxygen in all tissues of the body, both fluid and semi- fluid. . . . [There is] extreme saturation of the whole body with physically dissolved oxygen, so that the cells have a much higher reserve of oxygen than they normally have. . . . We may assume, then, that the increased amount of oxygen in solution provides a true reserve for the tissues and, consequently, that the tissue cells can withstand a circulatory arrest of longer duration." (7)

If a terminal patient could be kept in such a chamber, there would be a wider margin of safety when he died. Or if a newly dead patient were put in such a chamber, artificial respiration and heart massage might work more effectively.

With fully adequate preparation, equipment, and personnel, the cooling phase seems to present little problem in most cases. Heartlung machines and heat exchangers are available at many hospitals. The cardiopulmonary bypass technique is commonly used for open-heart surgery, with cooling of the blood and body from the normal of about 38C down to 20C, and sometimes lower; this technique has been described, for example, by Sealy and co-workers. (104) Apparently it could also be used, depending on the cause of death and opportunity for preparation, to cool freshly dead bodies quickly and safely, with no damage to the brain.

Finally, we ask how long it is safe to keep the body, after it has been cooled and before it has been frozen.

"......當一隻動物或是一個病人呼吸著三大氣壓 (絕對的) 的純氧時，在其體中所有的組織，將會有氧氣的物理性溶液的大增，不管是流體或是半流體......整個身體 [存在著] 以物理性溶解氧氣的極度飽和現象，因此細胞中有比一般狀況下要高出許多的氧氣儲存量......因此我們可以假設在溶液中所增加的氧氣量，會為組織提供一個真正的儲存，因而導致組織細胞可以承受較長時間的循環中止。" (7)

假使一個末期病人可以被置放在一個這樣的腔室之中，當他死後將會有較為廣大的安全空間。或是說一個剛逝世的病人假使將其置放在一個這樣的腔室之中，那麼人工呼吸和心臟按摩搞不好會更有效率地作用。

有了器材和人員完整足夠的準備，在大多數的案例中，冷卻的階段似乎都沒有出現什麼問題。在許多醫院中都配備有心肺機器和熱交換器。心肺呼吸系統繞道技術乃是非常普遍地用在開心手術，其間有將血液和身體從正常的38度冷卻到20度，而且有些時候還更低；這項技術已經有被說明過了，例如Sealy和其同僚所作的。(104) 依據死亡的成因和預先準備的機會而定，很明顯地這技術也可以用來快速地和安全地冷卻剛剛過往的身體，而不會有對腦度的傷害。

最後我們要問，在其被冷卻之後以及被冰凍之前，保留身體多少時間還會是安全的。

If a heart-lung machine has been used, and continues to be used, this time may be more or less indefinite.

If the brain has reached the vicinity of 10C without damage, for example by use of a heart-lung machine which then has to be disconnected, it can survive up to an hour without blood circulation, although there may be some relatively minor damage, if use is made of carotid arterial infusion of low molecular weight dextran; this is on the basis of experiments of Edmunds and co-workers with living dogs. (25)

Likewise the experience of Egerton and coworkers with patients undergoing hypothermic open-heart surgery showed that temperatures below 12C for more than forty-five minutes produced some brain damage, although most of them made complete recovery within four months. (26) Other work also shows that there is some brain damage, even if the blood still circulates, when the temperature reaches the vicinity of 0C, the freezing point of water.

Hence probably the body should not be cooled below about 10C before the freezing equipment is made ready, if this can be done within an hour or so later, as ought to be possible in a hospital.

Maximum and Optimum Storage Temperature

There seem to be four main possibilities for choice of storage temperature, and we must consider the theoretical and practical advantages and disadvantages of each. These are naturally occurring low temperatures in the arctic and antarctic, and the temperatures respectively of solid carbon dioxide, liquid nitrogen, and liquid helium.

假使有使用心肺機器，而且一直繼續用著，那麼這段時間可能或多或少是不定期限的。

假使腦部可以冷卻到攝氏10度的左右而沒有傷害，例如使用心肺機器，而且假使使用的方法乃是低分子量葡萄糖的頸首動脈灌輸，在其必需被移開之後，雖然可能會有一些相對少量的傷害產生，腦部可以在沒有血液循環之下存活到一個小時；此乃是依據Edmunds和其同僚針對活狗的實驗而得的。(25)

同樣地，Egerton和其同僚針對進行低溫開心手術病人研究的經驗中發現，在溫度低於攝氏12度下超過四十五分鐘，就會產生一些腦部的傷害，但是他們大部分都能在四個月內獲得完全的恢復。(26) 其它的研究也發現，當溫度達到水的冰點攝氏零度左右，就算血液還保持循環著，還是會產生一些腦部的傷害。

因此在冰凍設備準備齊全之前，身體可能不應該被冷卻到低於攝氏10度。假使這能夠在一小時之內或是稍微晚一點進行的話，應該盡可能在醫院中進行。

最大和最佳的儲存溫度

看起來似乎有四種主要的儲存溫度選擇的可能性，而我們必須要就理論上和實務上優點和缺點來考量每一個選擇。這些溫度就是在北極和南極自然有的低溫，以及分別是固態二氧化碳，液態氮和液態氦的三個溫度。

By way of general introduction, Dr. Audrey U. Smith says: "The basic principle in storing living cells is to arrest the processes of aging and degeneration. When living cells are cooled there is a slowing down of the biochemical processes involved in respiration, metabolism and all the other interactions between the cytoplasm of the cells and their environment. If they are cooled to temperatures in the range below -79C in which carbon dioxide and other gases are solidified or liquefied, all chemical changes must either be slowed to a minute fraction of the normal rate or else completely halted. Aging should not occur and it should be possible to preserve them for infinitely long periods in this temperature range." (110)

Of course, "infinitely long" is a slight exaggeration, and in fact we know that some kinds of cells stored at dry ice temperature, -79C, do show slow changes, with the percentage of living (revivable) cells decreasing week by week or even day by day, even though other kinds of cells have shown no appreciable deterioration after several years. For example, Meryman says: "In the case of blood frozen without glycerol significant decay is measured in days of storage at -70C, weeks at -80C, months at -90 C, and years at - 100C (70)

This does not necessarily mean that the relatively high temperatures are altogether hopeless. Some changes may take place, but little can yet be said about their extent and their reversibility. It may be that the changes, even though "fatal" by present tests, are minor, limited, and eventually reversible.

Audrey U. Smith博士以一般入門的方式說道: "儲存活細胞的基本原則乃是去將老化和變壞的過程中止。當細胞被冷卻之後，其和呼吸，新陳代謝和介於細胞的細胞質與其環境之間的所有其它的交互作用有關的生化過程就會慢下來。假使它們被冷卻到低於攝氏零下79度的溫度下，其中二氧化碳和其它的氣體就會被固化或是液化，所有的化學轉變一定不是被緩慢化到正常速率的一點點，就是被完全地中止。在這種溫度下，老化就不應該會發生，而且應該就可能來將其保存到一段無限期長的時間。" (110)

當然說 "無限期長的時間" 是有一點誇張，而且就算說其它種類的細胞在經過多年之後還是沒有顯現出可察覺的變壞，但是事實上我們知道有某些種類的細胞，儲存在乾冰溫度攝氏零下79度之中，還是會出現緩慢的改變，其活性 (可以復活的) 細胞的百分比會逐週地，甚或逐日地減少。例如Meryman所說: "就沒有使用甘油的冰凍血液案例而言，明顯的腐壞在攝氏零下70度乃是以日計的，零下80度乃是以週計的，零下90度乃是以月計的，而零下100度則乃是以年計的。" (70)

這不一定就意味著所有其它相對上較高的溫度就完全是沒有希望的。某些改變是可能會發生，但是目前還是很難描述它們的的程度以及它們的可逆度。也有可能這些改變雖然以目前的試驗來講是 "致命的"，但是其實是次要的，有限的，而且最後是可逆轉的。

It is not a case of general rot proceeding inexorably, although slowly; rather, it may be a case of some kinds of action not being completely inhibited, and stability may be reached after only some changes which, seen in perspective, are trivial.

Thus we cannot dismiss out of hand the suggestion sometimes heard that bodies be submitted to natural cold storage in arctic regions below the frost line. It has the obvious advantages of requiring no expensive investment and servicing, and of reduced vulnerability in event of war. However, the coldest natural temperature is well above that of dry ice, and probably too high. The odds would seem heavily adverse.

For extremely long-term storage, there seems to be nearly - but not quite - universal agreement that liquid helium temperatures, in the neighborhood of -270C, are safest. One of the dissenters is Dr. R. B. Gresham, who says: "It has been shown that after materials are frozen, there occurs a continuous thermodynamic activity down to -196C (-321F) or liquid nitrogen temperature, where movement ceases only to be noted again at -269C (-449F) or liquid helium temperatures.. . . . Although the effects of this thermodynamic activity on long-term storage of living cells is not known, when storage time is to be measured in years, it is theoretically desirable to maintain a temperature of -196C." (36)

This argument does not really seem very impressive. The "thermodynamic activity" and "movement" refer merely to certain irregularities in the rate of heat loss as the temperature is lowered, and accompanying shifts in the molecular structure or physical state of materials, mainly water.

這雖然是緩慢地，但是並非是一個全然嚴峻腐敗過程的情形，反而可能是某些種沒有被完全抑制的作用的情形，而且其穩定度僅要在經過某些改變之後就可能達到，這以透視的觀點來看是有點瑣碎。

因此我們不應該馬上對某些時候聽到的建議，也就是將身體置放在兩極區域低於冰凍線下的自然冰凍儲存，嗤之以鼻。這有不需要昂貴的投資和維護，而且在戰爭發生時，有降低受害危險度的一些明顯優勢。然而最低的自然溫度是大大地高於乾冰，而且可能還是太高。其成功概率看起來似乎是嚴重地不利。

針對極度長期的儲存而言，似乎有一個接近，雖然不全然是統一的共識，那就是液態氦的溫度，在攝氏零下 270 度左右是最安全的。有一個不同意的人，**R. B. Gresham** 博士表示，"實驗顯示物質在冰凍之後，仍然會有熱力學上的作用直到攝氏零下 **196** 度 (華氏零下 **321** 度)，也就是液態氮的溫度，而到攝氏零下 **269** 度 (華氏零下 **449** 度) 也就是液態氦的溫度，才能夠再度看到運動的停止．......雖然這種熱力學上的作用對活性細胞長期性的儲存上的後果還是不得知悉，但是如果儲存時間是要以年計的話，那麼理論上最好是去維持一個攝氏零下 **196** 度的溫度。" **(36)**

這個論點看起來真的似乎不太吸引人。這個 "熱力學上的作用" 和 "運動" 僅僅指出當溫度被降低時，在熱力喪失速率中的某些不規則現象，以及某些物質中，最主要的是水中的分子結構或是物理狀態上的附隨轉移。

As far as I can see, there is no particular reason to think this implies any instability at a fixed temperature, in general. Most writers do not seem to be worried by this question.

A more serious objection to use of the lowest temperatures is that while nothing will happen after storage temperature is reached, changes may take place en route. In other words, we should not use a temperature any lower than necessary, because we may be letting ourselves in for gratuitous trouble. In every range, more cooling means more change, and unnecessary changes are to be avoided.

On a practical level, liquid helium is relatively expensive, and tricky to handle.

What would seem to emerge, then, is the following. At the present time, the temperature of choice is that of liquid nitrogen. When permanent installations are built, probably liquid helium will be used. As an emergency or austerity measure, one might use dry ice, which is cheap and easy to handle.

Radiation Hazard

Would a body in cold storage, although preserved against decay, be gradually "cooked" by the slow but inexorable attack of natural radiations?

就我所能看到的,沒有什麼特別的理由來認為在一個固定的溫度下,總括而言這會意含著任何的不穩定度。大部分的著作者看起來似乎都不會去煩惱這個問題。

一個反對使用最低溫度較為嚴重的理由乃是雖然在儲存溫度已經達到後,將不會有任何事情發生,然而改變還是可能會沿途發生。換句話說,我們不應該去使用一個比需要上還要低的一個溫度,因為我們可能會讓我們自己招惹莫名其妙的麻煩。在每一個範圍中,越多的冷卻就意味著越多的改變,而這些莫需有的改變是應該去避免的。

以務實的觀點來看,液態氦不僅是相對上較昂貴,而且在操作上也比較麻煩。

這樣看起來會浮現檯面的是如下的情形。在目前,溫度的選擇乃是液態氮的溫度。當永久性的設施建造完成之後,可能將會使用液態氦。用在緊急或是保守的狀況下,一個人可能會去使用乾冰,因為它不僅較便宜而且較容易操作。

輻射的危險

當一個身體處於冰凍儲存狀態,雖然是被保護來對抗腐敗,但是是否會被那些雖緩慢卻不間歇的自然輻射線的攻擊,而逐漸地被 "煮熟" 呢?

We know these are all around us: cosmic rays bombard us from the skies; uranium, thorium, and radium in rocks and soil, in concrete and brick, spray penetrating emanations similar to X rays; and certain radioactive atoms (radioisotopes) in our own bodies dribble slow poison. (In addition to this natural "background" radiation, there is fallout radiation from testing of nuclear weapons, but this so far is more or less negligible.)

Since these radiations are of low intensity, they produce only a "chronic dose" which is scarcely noticed, since a functioning body can repair most of the damage as fast as it occurs. But all doses absorbed by a body in cold storage must be regarded as acute; we must consider the possibility that the cumulative damage to a frozen body might become serious as the centuries passed.

Examining the data, we find there may indeed be a problem, but not one too formidable. (The pertinent information can be found, for example, in The Effects of Nuclear Weapons, U. S. Atomic Energy Commission, 1962.)

The unit usually used to measure dosage of radiation is the "rem" (roentgen equivalent, mammal, or man); we do not need its technical definition, but may note that an acute dose of 100 rems or less is unlikely to produce noticeable illness, a dose of 600 rems results in severe radiation sickness requiring hospitalization and the most competent care, and a dose of 1000 rems or more is almost certainly fatal with the present resources of medicine.

我們知道這些都在我們的週遭的：宇宙射線從天空轟擊我們；在石頭和土壤中，或是在混凝土和磚塊中的鈾，鍶和鐳，會像 X 射線一樣，噴灑出具穿透力的的放射物；而在我們自己身體裡面的某些放射性原子 (放射性同位素) 會逐漸滴出慢性毒藥。(除了這個自然的 "背景" 輻射之外，還有從核子武器試驗而來的落塵輻射，但是這項目前或多或少是可以忽略的。)

既然這些輻射乃是屬於低強度的，所以它們所產生的僅是一種 "慢性劑量"，一般幾乎都察覺不到，因為一個正常運作的身體可以在傷害一產生時，就馬上將其大部分修復。但是在冰凍儲存中的身體全部所吸收到的劑量，應該要被認定為嚴重的；我們必須要去考慮到在一個冰凍身體上的累積傷害，當幾世紀經過後，可能會變成相當嚴重的可能性。

審視過科學數據，我們發現這的確會是一個問題，但是沒有什麼可太驚懼的。(相關的資訊可以在例如 The Effects of Nuclear Weapons, U. S. Atomic Energy Commission, 1962 中找到。)

通常用來量度輻射劑量單位的是 "rem" (Roentgen當量，人或是哺乳動物)；雖然我們不需要去知道其技術上的定義，但是可能應該要知道一個100 rem 或較少的急性劑量，是不太可能會產生可察覺的疾病，一個600 rem 的劑量就會導致嚴重的輻射疾病，而需要住院和最專業的照料，而一個1000 rem 或更高的劑量，以目前所有的醫療資源來講，這幾乎一定是會致命的。

Background radiation varies considerably with locale, but as a rough average we might expect everyone to receive a dose of about 10 rems in 50 years. A stored body, then, might take 500 years to accumulate a "clinical" or symptom-producing dose of 100 rems, and 3,000 years to soak up a currently dangerous dose of 600 rems. These times, to be sure, might be reduced if nuclear war or excessive weapons testing produced heavy fallout; but they could also be much lengthened by precautions available at moderate expense.

If the bodies were stored underground, in vaults made of low-radioactivity materials, then they would be shielded from most of the external background radiation, leaving only the internal radiators to worry about. These consist mainly of one of the forms of the element potassium (the radioisotope potassium-40) found especially in the soft tissues of the body.

The dose rate due to potassium-40 is about 20 millirems (0.020 rem) yearly. This will continue essentially indefinitely, since the "half life," or time required for the dose rate to halve as the decaying potassium-40 is used up, is over a billion years. But to accumulate a dose of 100 rems would take 5,000 years, and for 600 rems the wait would be 30,000 years.

Even then, the radiation damage would no doubt be substantially less than the injury done (to the earliest frozen bodies) by crude freezing methods, so one might guess that at least 100,000 years must elapse before radiation damage becomes critical.

背景輻射雖然是會隨著地點會有相當大的變異，但是就一個粗略的平均值而言，我們可能可以預期每一個人在五十年內，會接受到大約 10 個 rem 的劑量。因此，一個儲存起來的身體可能需要 500 年來累積一個 "醫療上" 或是病症產生的 100 rem 劑量，而需要 3,000 年來吸收足一個目前認為危險的 600 rem 劑量。確切地說，假使核子戰爭或是過量的武器試爆產生出濃重的落塵，上述這些時間可能就會被縮短；然而它們也有可能藉著所有花費中度的預防措施而得以大大地增長。

　　假使身體是被儲存在地下，而且是置於用低放射性的材料所製成的容器內，那麼他們就可以防護掉大部分外來的背景輻射，而僅留下內部的輻射源需要擔心。這些主要地是包含元素鉀的各種型態之一 (鉀–40 的放射性同位素)，特別是在身體的軟性組織中常見的。

　　鉀–40 的劑量率大約是每年 20 millirem (0.020 rem)。這劑量率實質上是會一直繼續到無限期的，因為其 "半衰期"，或是當蛻變中的鉀–40 被耗掉到其劑量率變成一半所需要的時間乃是超過十億年。但是要累積到一個 100 rem 的劑量就需要 5,000 年，而要達到 600 rem 的劑量則需要等上 30,000 年。

　　就算這樣，其輻射傷害毫無疑問地還是大大地少於因為粗糙的冰凍方法 (對最早期的冰凍身體) 所造成的損傷，所以我們可以猜測至少一定要經過100,000年之後，輻射傷害才有可能會變成是重要的。

I can think of certain heroic measures to extend this time to a million years or more, but it is hardly worth the trouble.

Most of us will be frozen by advanced methods developed in the next decade or two, and will be waiting in cold storage mainly for the solution of the aging problem. In light of the explosive acceleration of scientific progress, it would be astonishing if this were to take as long as 5,000 years. In this view, we can ignore the effects of bodily radiation damage.

However, a postscript may be worthwhile to reassure those worried about the genetic effects of radiation. It is true that a dose of 100-300 rems inflicted on everyone in every generation might eventually produce so many mutations or freaks of inheritance as to threaten the race, if nothing were done about it. But we expect eventually to control and tailor our genes, the physical blueprints of inheritance carried by our cells, and in any event the resurrected frozen will not constitute the entire populace. Individually, there is no cause for concern: a man exposed to 500 rems has only a negligible chance of observing deformities in his children or grandchildren. (See, for example, the article by Professor Muller in Radiation Biology, ed. Alexander Hollaender, McGraw-Hill, 1954.)

我可以想出一些大膽的方法來將此時間延長到一百萬年或是更長，但是這幾乎是不值得去耗費這種精力的。

　　我們大部分的人將會利用未來十年或是二十年中所開發出來的先進技術來冰凍，而且在冰凍儲存中等待的目的主要是老化問題的解決。看到科學進展爆發性的加速，假使這還需要長達 **5,000** 年的時間是會令人訝異的。就此觀點，我們就可以來忽略掉人體輻射傷害上的效應了。

　　然而為了要來讓那些擔心輻射上的基因效應的人放心，可能值得我們做一個後記。的確，假使我們對此沒有任何作為的話，一個**100–300 rem**的劑量加諸在每一代的每一個人，到最後可能會產生出許多的遺傳上突變或是怪物，以至於威脅到整個人種。然而我們預期最後將可以來控制和修改我們的基因，也就是由我們細胞中所攜帶的遺傳上的實質藍圖，況且在任何狀況下，被復甦的冰凍者也將不會包含整個人口。就個人而言，也沒有什麼好擔心的：一個暴露到**500 rem**劑量的人，要在其孩子或是孫子中觀察到畸形的機率可以說是小到可以忽略的。(參考例如**Muller**教授在其輻射生物學一書中的文章，由 **Alexander Hollaender**編輯，**McGraw-Hill**出版，**1954**。)

CHAPTER III

Repair and Rejuvenation

修復和回春

CHAPTER III

Repair and Rejuvenation

We have investigated the proposition that a freshly dead body can be frozen, stored for a long time at low temperature, and thawed again without excessive damage. But after this has been done, it is still just a freshly dead body (although it may be hundreds of years old) and much more is required.
We need assurance that we can be revived, and not only that; if we die sick, we want to be made well; if we die broken, we want to be made whole; and if we die old, we want to be made young.

(Indeed, we want yet more. We hope to be made not only good as new, but eventually much better than new. This part of the discussion, however, will be mostly reserved for later chapters.)

Absolute, rigorous proof of the future capabilities of science cannot, of course, be offered. For example, no engineer today could prove that it will ever be possible to manufacture a cheap, safe, reliable family helicopter. He cannot prove it can be done, because he does not know exactly how to do it.Nevertheless many engineers, and probably most, would make such a prediction with confidence; there are favorable hints in current research, and the whole march of history is obviously in this direction.

第三章

修復和回春

我們已經探討過一個剛逝世的身體，可以被冰凍儲存在低溫下一段長久的時間，然後被解凍而沒有產生過度的傷害的一些見解。但是在此完成之後，他畢竟還僅是一具新鮮的屍體而已（雖然他可能已經是好幾百歲了），需要做的事情還多得很。我們需要去確認我們真的可以被復甦；而且不僅如此，假使我們是因病而死的，我們必須要能夠被治癒；假使我們是因破壞而死的，我們需要被恢復完整；而假使我們因老化而死的，我們則需要能夠被變成年輕。

（的確，我們要的是越多越好。我們希望被變得不僅要和新的一樣好，而且最後要比新的還要好。但是有關這一部份，讓我們將其大部分留在後續的章節中再討論。）

有關絕對地和嚴謹地來證明未來科學的能耐，當然是不可能被給予的。例如，當今的任何一位工程師都不可能證明未來將有可能製造出一架又便宜，又安全，又可靠的家庭用直昇機。他不可能證明這是可行的，因為他目前不知道真正地如何來完成。然而有許多工程師，也有可能是大多數的工程師，都會有信心來做這樣的一個預測。在當今的研究結果中，充滿了各種有力的證據，而且整個歷史文明的步伐，很顯然地也都是朝著這個方向在前進。

That future technicians will be able to repair and rejuvenate us cannot be proven, but it can be made convincing all the same. Let us review briefly, and with no strenuous effort to be orderly, some of the striking accomplishments and prospects of modern medicine and biology, especially those relevant to repair and rejuvenation.

Revival after Clinical Death

It is well known that hundreds of people have been revived after being clinically dead for many minutes, i.e. after heartbeat and breathing stopped.   Most of these died of such things as heart trouble, shock, asphyxiation, and drowning. They were revived by rather simple measures including artificial respiration, blood transfusions, heart massage, and stimulation by drugs or electricity. (97)

A remarkable example of what can be done even in these primitive times is offered by the case of Professor Lev Landau, a famous Russian physicist who suffered an auto accident in 1962. He is reported to have sustained a fractured skull, brain contusions, severe shock, nine broken ribs, a punctured chest, a fractured pelvis, a ruptured bladder, paralysis of the left arm, partial paralysis of the right arm and both legs, and failing respiration and circulation. During the next fourteen months he died four times, and was four times resuscitated. In the spring of 1963 he was still alive and apparently improving. (180)

The people in the freezers -- you and I -- will have died mostly of disease or old age. The immediate cause of death will usually be the failure of some vital organ. The future medicos, then, will perhaps proceed somewhat as follows:

雖然未來的技術人員將有辦法修復和回春我們，這是無法被證明的，但是大家一致都相信這是可以被達成的。現在讓我們不特別費功夫地拘泥在秩序上，來簡要地回顧一下當代醫學和生物學上的一些令人訝異的成就和前瞻，尤其是在那些和修復以及回春有關的方面。

醫療死亡後的復活

　　大家都知道，有好幾百人都在醫學上死亡後，也就是說在心跳和呼吸中止好幾分鐘之後再度被救活。這些大部分都是死於心臟問題，休克，窒息和溺水諸類的情形。他們都是用頗簡單的方法來被救活的，其中包括了人工呼吸，輸血，心臟按摩和用藥物或是電流的刺激等。　**(97)**

　　一個著名的俄國醫生 **Lex Landau** 教授，在他 **1962** 年時慘遭一次車禍的案例中，提供了一個就算處在這麼原始的時代裡，我們還能做些什麼的絕佳例子。依據報告，他遭受到頭顱破碎，腦部瘀血，嚴重休克，肋骨九根斷裂，胸部刺傷，骨盆破裂，膀胱破損，左手臂麻痺，右手和雙腳部分麻痺，以及呼吸和循環的失敗。在後來的十四個月中，他曾經死了四次，而也被救活了四次。到 **1963** 年的春天，他仍然活著，而且很顯然的還在進步中。**(180)**

　　你和我，這些在冰凍櫃中的人，大部分都將會是死於疾病或是年老。而最直接的死亡原因，通常將會是某些重要器官的失敗。因此未來的醫生可能會進行類似下面的一些事情：

first, either restore or provide respiration and circulation; next, repair or replace the defective organ that was the proximate cause of death; then cure any acute disease and make any other urgent repairs; lastly, and at leisure, make a general overhaul and rejuvenation.

The first stage, the restoration and support of life while biological repairs are being made, will demand the use of mechanical devices, some of which are already well known.

Mechanical Aids and Prostheses

There exists right now a rather imposing list of inventions to perform biological functions. To assist breathing, we have respirators of various kinds, oxygen masks, pressure chambers, and iron lungs. To help a faulty heart keep proper time we have electronic "pacemakers," some of which can be implanted in the body. There are pumps which can be connected to the circulatory system and do the work of the heart. There are even machines which can take the place of both heart and lungs, aerating the blood as well as pumping it. These are all matters of common knowledge.

Slightly less familiar is the use of a machine to perform the function of a missing or diseased kidney. Dr. B. H. Scribner of the University of Washington, for example, has been reported treating patients once or twice a week by running their blood (out an artery and back into a vein) through the device, which removes the wastes ordinarily handled by the kidney.

首先，不是去恢復不然就是去提供呼吸和循環； 接著，去修復或是更換那個導致死亡最接近原因的失靈器官； 然後治療任何急性的疾病和進行任何其它緊急的修理； 最後，並且是輕鬆地，進行一個整體的檢修和回春。

第一個階段，當在進行生物性修復時，生命的恢復和維持將會需要去使用一些機械性的器材，而其中某些東西已經是大家都耳熟能詳的。

機械輔助和器官替換

現在已經有一張看起來頗驚人的發明清單，可以用來執行生物性的功能。在輔助呼吸上，我們有各種不同種類的呼吸器。氧氣面罩，壓力艙和鐵肺等。在幫助一個有瑕疵的心臟來保持其恰當的律動，我們有電子的 "心律調整器，pacemakers"，其中某些還是可以被植入體內的。還有一些幫浦可以和循環系統連結，而來代替心臟的功能。甚至於還有一些機器可以用來替換心臟和肺臟的地位，同時也可以在幫浦時氧化血液。這些已經都是家喻戶曉的東西了。

大家較不熟悉的就是使用一台機器，來執行一個少掉的或是病壞的腎臟的功能。例如依據報導，華盛頓大學B. H. Scribner博士，藉著將病人的血液打過這種機器 (從動脈打出，然後再打入靜脈)，連續地對他們進行一個星期一次或是一個星期兩次的治療，這樣就可以移除掉一些本來應該由該腎臟處理的廢物。

Just a very few years ago these patients would have been doomed, unless they were lucky enough to get a successful kidney transplant, but now they can apparently live indefinitely without kidneys. (21)

Pacemakers or electronic stimulators have been used not only for the heart, but also in cases (as after abdominal surgery) when the intestine is paralyzed, and in the case of dogs when the bladder is paralyzed. (82)

The future outlook for prosthetics is even more impressive. Dr. Lee B. Lusted (professor of biochemical engineering, University of Rochester) thinks that within fifty years it will be possible to replace nearly all of the body organs by compact artificial organs with built-in electronic control system -- including, for example, the heart, kidneys, stomach, and even the liver. (64) (Imagine an artificial stomach, which would tolerate unlimited insult in the way of greasy and spicy foods, and never dream of growing an ulcer! Imagine a gin-resistant liver! Like most blessings, these would no doubt be mixed.)

Artificial limbs are less important than vital organs, but tremendously advanced arms and legs will be available from the shop if needed. Russians at the Central Scientific Research Institute for Prosthetics in Moscow already claim they have an artificial hand device operated by thought control! A metal bracket strapped to the arm is supposed to pick up biopotentials (electrical nerve impulses) generated by an effort of will; in other words, they claim the nerves of the body are used to control metal instead of muscle. Furthermore, they say they are working on a way to produce artificial hands with a sense of touch! (11)

剛剛在沒有多少年之前，除非他們非常幸運到可以獲得一次成功的腎臟移植，否則這些病人一定會是已經命喪黃泉了。然而現在很顯然地在沒有腎臟之下，他們還可以不限期地活著。**(21)**

心律調整器或是電子刺激器不僅是可以用在心臟，同時也可以用在 (如在腹部手術後) 當腸道變成麻痺的一些案例，以及用在當狗的膀胱變成麻痺的案例。**(82)**

器官部分替換的未來前景則更是令人讚嘆。**Lee B. Lusted** 博士 **(Rochester** 大學生化工程教授) 認為在未來的五十年內，用有內建電子控制系統的精密人工器官，來替換身體裡面幾乎所有的器官將會是可能的，其中包括例如是心臟，腎臟，胃，以及甚至是肝臟。**(64)** (試想一個人工胃，它可以無限度地忍受油膩和辛辣形式的折磨，而且根本不需要去擔心會罹患胃潰瘍！ 試想一個不畏酒精的肝臟！ 但是就如同大部分的福氣一樣，無疑地，這也會是福禍參半的。)

雖然人工肢體會比各種主要的器官要來得較不重要，但是如果需要的話，未來在商店裡面將就可以買到異常先進的手臂和腿腳。在莫斯科中央器官替換研究中心的一些俄國人宣稱，他們已經開發出一種用意念來控制的人工機器手臂！其中有一塊綁在手臂上的金屬包片，乃是用來擷取由一個意志行為所產生的生物電位 (神經脈衝電力)；換句話說，他們認為身體中的神經乃是用來控制心靈，而不是用來控制肌肉的。還有他們表示，他們現在正在研究一種方法可以製造出具有觸感的人工手臂！ **(11)**

Since there have been many sensational announcements from Russia which turned out to be exaggerated or premature, a healthy skepticism is proper in this instance. Nevertheless, the principle is sound, and sooner or later the hardware will be ready. Mechanical limbs and their controls are so far crude, bulky, and inefficient, but steady strides are being made toward miniaturization of both controls and motors, with only power sources still lagging behind from the point of view of compactness. To get an idea of how compact computing machinery used for controls may become, we need only note what Dr. Fernandez-Moran has said: "At the present   time....   advanced   techniques   of ultraminiaturization   . . are bringing us ever closer to practical realization of information storage and integrated electronic circuits at the molecular level ..." (31) A computing machine which functioned at the molecular level might rival the human brain for   compactness! And to get an idea of how small machinery may be, we may   note that in 1961 a young engineer collected a $1,000 prize by constructing an electric motor .oo6 inch in diameter. (126)

Artificial organs and limbs will be used if natural repairs or replacements cannot be had; and they will probably be used again in the more distant future when they become so efficient that they are to be preferred over the biological. But our ability to repair or replace natural organs and limbs is growing rapidly, and probably for a long time this will be the dominant theme.

因為從俄國有發出太多挑動人心的訊息，而其實都只是誇張之詞，或是言之過早的東西，所以在這種狀況下，保有一個健康性的懷疑乃是合理的。但是儘管如此，其基礎原理仍然是穩固的，所以遲早其硬體就會被製造出來。目前機械四肢和其控制雖然是粗糙，笨重而且是沒效率的，但是在朝向控制和馬達的微形化上，一直都還是保持著穩定前進的腳步，只有電源方面，如果從精密度的觀點而言，還是有點落後。要能夠知道用來控制的電算機器將會變得多小，我們僅要去注意 Fernandez-Moran 博士所說的: "在目前.......超級微形化的先進技術........正在將我們帶向越來越朝分子尺度上的資訊儲存和積體電路的實質實現........。" (31) 在分子尺度上運作的電算機器中，其體積之小將可能可以媲美人腦！而要去想像機器將可以變成多小，我們可以去注意，在 1961 年就有一個年輕的工程師，他以建造出一個直徑 0.006 英吋的電動馬達而贏得了美金一千元的獎金。(126)

假如自然的修復或是替代已經是不可能時，人工的器官和四肢就可以被用到。而且在較遠的未來，當它們的效能勝過自然生物性的，而被大家偏愛時，它們就可能會再度地被重用。然而，我們修復或是替換自然器官和四肢的能力也正在快速地進步，而且可能會有一段很長的時間，這將是主要的研發題目。

Transplants

The routine transplanting or grafting of any organ of the body (obtained for example from a fresh corpse or from a cold-storage bank) is not yet possible, because the "immune reaction causes the body of the host to reject the "foreign matter." But leading biologists such as Dr. Jean Rostand are confident this barrier will be overcome. (119) In fact, it has been partly overcome already, and very impressive advances have also been made in the surgical techniques needed for these operations.

One of the very difficult but very active fields is that of lung transplants. Apparently the first successful reimplantation (into the same animal) of a lung taken from a dog occurred in 1951. In 1963 Dr. S. L. Nigro and coworkers reported this technique so far perfected that dogs are surviving after 1 1/2 years on the replanted lung alone. (83) Dr. J. D. Hardy of the University of Mississippi, a leading worker in the field, in 1963 reported temporarily successful transplants of lungs from one dog to another; in some cases he kept the lungs in cold storage for two to six hours before using them. The immune reaction, however, was not sufficiently suppressed, and the transplanted lungs eventually died. (38)

Dr. Hardy was also reported in 1963 to have performed the first human lung transplant. A fifty-six-year old man, a heavy cigarette smoker, had cancer of the left lung as well as impairment of the right lung. The donor lung was taken from a man of about the same age immediately following death. After the operation, the patient was reported as doing well. (24)

器官移植

身體上任何器官 (例如從一個新鮮的屍體或是從一個冰凍儲存庫來的) 例行的移植或是轉接，雖然目前還不是可能，因為接受者身體中的 "免疫反應" 會對這個 "外來物體" 產生抗拒。但是，像是 Jean Rostand 博士這類領先的生物學家都深信這種障礙將會被克服。(119) 事實上，這已經有被部分地克服了，而且在這類手術上所需的外科技術，也已經有非常驚人的進展。

其中有一個非常困難但是卻是非常活躍的領域，那就是肺臟的移植。很明顯地，第一次從一隻狗上取出肺臟，然後成功地再度植入 (植入同一隻動物)，乃是發生在 1951 年。而在 1963 年，S. L. Nigro 博士和其同僚發布，在只有肺臟再度植入之下，這種技術目前已經進步到在經過移植一年半之後，這些狗還能活著。(83) 密西西比大學的 J. D. Hardy 博士是在這一領域的領先科學家，他於 1963 年發表出從一隻狗到另外一隻狗的短暫成功的肺臟移植； 在其中的某些案例裡，在使用肺臟之前，他曾經先將其冷藏到二到六個小時。然而，由於免疫反應沒有被充分地抑制，所以這些移植的肺臟到最後都死掉了。(38)

Hardy博士在1963年也被報導進行了第一次人類肺臟的移植。一個五十六歲煙癮很重的老人，其左肺已經罹患癌症，右肺也遭破壞。捐出的肺臟乃是從一個大約同年齡的男人，在其死後馬上被取出而來的。據報導在手術之後，病人的狀況良好。(24)

It is not clear from this report, however, whether new methods were used to suppress the immune reaction or whether only temporary success is expected.

Another major organ, the kidney, has often been successfully transplanted. In earlier years there was usually very limited success unless the donor was a twin of the host; in this case, since they share the same genetic heritage, the new kidney knows the password, so to speak, and is not shot down as an intruder. Recently, however, the use of certain drugs and of X-ray treatment to suppress the immune reaction has led to more success with transplants from more distant relatives or strangers.

Many other examples could be cited. A recent one, rather interesting if not tremendously important, concerns tooth transplants reportedly made by Dr. Miklos Cserepfalvi of Washington, D.C. Of 146 tooth transplants made since 1956, 140 resulted in permanent, live teeth; by contrast, earlier attempts had usually resulted in rejection of the new tooth as foreign matter within a year or so. The teeth had been extracted from children eight to twelve during orthodontic work; they had not yet erupted through the gum, and the tooth was removed complete with the sac surrounding it. Dr. Cserepfalvi is quoted as saying, "There is no reason for anyone in this country today to have a false or missing tooth in his head." (15)

但是從此報導中，不太清楚他是否有使用新方法來抑制免疫反應，還是是否其成功僅僅是短暫的。

　　另外一個主要的器官，腎臟，也常常被成功地移植。在其早期年代裡，除非捐贈者是受贈者的雙胞胎，否則其成功通常是非常有限的。在雙胞的案例裡，因為它們都有同樣的基因遺傳，也就是所謂這個新的腎臟知道通關密碼，所以不會被當作是入侵者而被打垮。但是最近，因著使用某些藥物和 X-光療法來抑制免疫反應，導致在從更遠的親戚間，甚或是陌生人間來的移植有更高的成功率。

　　還有很多其它的例子可以被引述。最近的一個，雖然不是非常地重要，但是卻也頗有趣的，具報告乃是有關華盛頓特區的 Miklos Cserepfalvi 博士所完成的牙齒移植。自從其 1956 年以來所進行過的 146 個牙齒移植中，有 140 個移植都變成永久活性的牙齒。相對之下，大約在一年之前的嚐試，通常都導致把新牙當作外來物體般地排斥。這些牙齒乃是從八歲到十二歲小孩，在其進行牙齒矯正時所拔取而來的。它們都還沒有從牙齦中突長出來的，而且這些牙齒乃是連著其週遭的胞囊一起被移除出來。引述 Cserepfalvi 博士所說的，"當今，在這個國家裡，沒有任何一個人有任何理由來在其頭顱中，有一顆假牙或是一顆空牙。" (15)

The general outlook is entirely favorable. In 1963 Dr. Robert Brittain (University of Colorado) and Dr. Richard Lillehei (University of Minnesota) are reported to have said at a convention of the American College of Physicians that within only five years it will be possible successfully to transplant all human organs, except those of the central nervous system! (40)

But even the central nervous system is not inaccessible to these techniques, although mastery will take longer, and of course we are only interested in repair of the brain, and not its replacement. A Yugoslav researcher, Dr. Mira Pavlovic, has successfully grafted a large part of the brain of one embryo chick to that of another; some of these subjects hatched and lived as long as two months. (119)

As already suggested, the organs for transplant will often be obtained from cold storage banks. Dr. Lillehei, together with Drs. Bloch and Longerbeam, expect surgical deep freezes of the near future to store kidneys, spleens, and lungs, as well as other organs. They have already quick frozen organs, kept them up to two weeks at dry ice temperature, then thawed them rapidly by microwave diathermy, treated them with a substance called LMD (low molecular weight dextran) and replanted them. (102)

But now a question arises: if everybody is frozen, there will be no cadavers to scavenge; where will the spare parts come from? Fortunately, answers are in sight.

整體的遠景看來是完全地樂觀的。據報導在 1963 年，科羅拉多大學的 Robert Brittain 博士和明尼蘇達大學的 Richard Lillehei 博士，在一個美國醫師學院的大會中表示，僅要在五年之內，除了中樞神經系統之外，成功地移植人類所有的器官將會是可能的！ (40)

　　雖然技術的成熟將會要耗時久一點，但是就算中樞神經系統，也不是這些技術不能接近的禁臠之地，而且我們所想要的當然僅是大腦的修復，而不是大腦的移植。一個南斯拉夫的科學家 Mira Pavlovic 博士已經成功地將一隻胚胎雞大腦的一大部分轉植到另外一隻上面，其中有些試驗體都能夠成功孵出，而且還活到兩個月長的時間。(119)

　　就如同我們提過的，這些用來移植的器官，通常將會是從冰凍儲存庫中取得的。Lillehei 博士和 Bloch 博士以及 Longerbeam 博士都預期，在未來的手術上，深度冰凍將會去儲存腎臟，脾臟和肺臟，以及其它的器官。他們已經快速冰凍過一些器官，在乾冰溫度下儲存了高達兩個星期，然後藉著微波透熱法將之快速解凍，用一種稱為 LMD (low molecular weight dextran) 的東西處理過它們之後，接著再度將它們植入。(102)

　　但是現在出現了一個問題：假使每一個人都被冰凍起來了，那麼就沒有屍體可以來取材了； 這樣器官將從哪裡來呢？很慶幸的，答案已經可以看到了。

In the relatively near future, and for a certain period in history, we may use organs from lower animals. There has been some success in suppressing the "immune reaction" even in the case of "heterografts," or transplants between different species.

On December 22, 1963, many newspapers throughout the country featured the remarkable story of Jefferson Davis, a New Orleans dock worker, whose diseased kidneys were replaced by those of a ninety-pound chimpanzee. The historic operation was reported performed by a team of Tulane University surgeons headed by Dr. Keith Reemtsma. A few days after the operation, the grafted kidneys seemed to be functioning satisfactorily, although the prognosis, of course, was uncertain. It seems unlikely that this pioneering effort will be completely successful. The measures now used to induce the host body to tolerate the foreign tissue are usually inadequate, unless applied in such massive doses that the side effects become critical. Radiation and/or presently known drugs have poisonous effects themselves; furthermore, along with suppressing the immune reaction, they depress the body's ability to fight off infections, opening the way to complications. But the research is being vigorously pressed, and it may not be many years before lower animals can supply us with replacements for such organs as the lungs, kidneys, heart, liver, spleen, stomach, or pancreas.

在相對上較近的未來，而且在歷史上的某一段時期中，我們可能會用來自較低級動物的器官。在抑制"免疫反應"上我們已經有了一些成就，甚至於在"異類轉植"的案例，也就是在不同種間的轉植中也有。

在 1963 年 12 月 22 日，全國的許多報紙都報導了一個有關紐奧良碼頭工人 Jefferson Davis 的故事，他病壞的腎臟被用一隻九十磅重猩猩的腎臟替代了。據報導，這個歷史性的手術乃是由 Keith Reemtsma 博士所領導的一個 Tulane 大學的外科醫生團隊所進行的。手術後幾天之後，這個轉植的腎臟看起來似乎運轉得相當令人滿意，雖然未來會是如何當然是不可確定的。這個先驅性的努力看起來是不可能會是完全地成功的。目前用來誘使接受者的身體，來忍受外來組織的方法，通常都是不足的，除非其所用的劑量多到其副作用變成關鍵因素。輻射和／或是現今所知藥物本身，都具有毒性效應；況且，隨著免疫反應的抑制，它們也會降低身體對抗感染的能力，因而打開併發症的大門。然而這方面的研究現在正如火如荼地進行之中，因而可能不到幾年之前，較低等動物就可以供應我們一些替代器官，例如肺臟，腎臟，心臟，肝臟，脾臟，胃，或是胰臟。

Organ Culture and Regeneration

To the question of where the spare parts will come from in the more distant future, there is a beautifully simple answer: they will come from ourselves!

We know that germ cells - sperm or egg - produced by our reproductive organs contain chromosomes carrying our genetic information or blueprints, and that these germ cells, after combining with one from the opposite sex, are capable of developing into complete human beings. It is less widely known among laymen that either the sperm or the egg alone is considered capable of developing into a person, although so far this has seldom if ever happened. (119) Still less is it realized that ordinary body or somatic cells, which also carry the chromosomes, may retain potential "totipotence"; even though they are differentiated and specialized, it may be possible to reverse and then generalize their development, and in fact such cases have actually occurred, where an ordinary body cell (in certain lower forms of life) has taken the place of a germ cell and led to growth of a complete individual. (79)

The possibilities, then, are obvious. As soon as enough is known about guiding growth and development, a germ cell or an ordinary somatic cell, perhaps even from the skin, can be taken from the resuscitee's body, and from this will be grown, not a complete individual, but just the organ or organs needed for repair.

器官培養和再生

有關在較遠的未來，預備的器官部件要從哪裡來的問題，其實有一個簡單且漂亮的答案: 它們將會從我們自己的身體而來！

我們知道由我們的生殖器官所製造出來的生殖細胞，精子或是卵子，都包含著攜帶我們的基因資訊或是藍圖的染色體，而這些生殖細胞在和一個從異性而來的結合之後，就有可能成長成一個完整的人類。較不為一般外行人所普遍知悉地，光是精子或是卵子單獨地就有辦法來成長成一個人，雖然如此，這種事情到目前為止，如果有也是少之又少。(119) 而這事情以身體細胞或是所謂的體細胞的方式實現，則就更是少見了。體細胞也攜帶著染色體，雖然它們已經被分類化和專一化了，但是可能還保留著 "全潛能" 的能力。 它們可能是可以被逆轉，然後來一般化它們的轉化，而且事實上這種案例已經真正發生過了，其中有普通的體細胞 (在某些較低等形式的生命中的) 替代了一個生殖細胞，而來導致一個完整個體的長成。(79)

因此，這些可能性是非常明顯的。當有關成長和轉化的導引被足夠地理解之後，從被復甦者的身體中所取出的一個生殖細胞，或是一個普通的體細胞，甚至於可能是從皮膚取來的細胞，就可以從其中培植成長出僅是用在修復上所需要的單一器官或是各種器官，而不需要是一個完整的個體。

It will not even be necessary to suppress the immune reaction when making the implant, because it will be his own tissue; it will be an autograft and not a homograft. You may get a new heart, for example - your very own, exactly like the original, but young and strong and ready for another three-score-ten of faithful pumping service.

Can we really be sure this will come about? Is not the guidance of growth and development an exceedingly intricate and difficult business? The answer, as usual, is that the problem is indeed complex, but there is ample optimism among the experts and there are already successful beginnings.

Tissue culture, the growth of cells in a test tube or other artificial environment, is of course old hat. The famous Dr. Carrel "maintained a strain of chick embryonic cells in this way for more than thirty years (much longer than the lifetime of the hen which would have grown from the embryo). . . ." For obvious reasons this strain of cells was called Carrel's 'immortal' strain. It died (of neglect) during the Second World War... (87) Complete organs have also been maintained outside the body of the animal for varying lengths of time, and test-tube grown organs require no great stretch of the imagination.

Speaking in a somewhat different but nevertheless appropriate sense, Dr. Philip Siekevitz (Rockefeller Institute of Medical Research) has said, "I shall not be surprised if in our lifetime we know in general, often in specific, terms how the body regulates its growth. And to know is to influence." (105)

在進行移植時，甚至就可以不需要去抑制免疫反應了，因為這將是他自己本身的組織；這將是一個自體中的轉植，而不是一個異體間的轉植。例如，你將可能獲得一個真正屬於你自己的一個新的心臟，跟你本來的完全一樣，然而卻是既年輕而且又強壯的，而且又可以忠實地提供另外一個七十年的輸血服務。

我們能否真正地確認這將成真呢？成長和轉化的導引，不是一種極度地精密而且是困難的事情嗎？還是一樣，這個答案乃是，雖然這個問題的確是非常地複雜，但是在許多專家中，還是充滿了極大的樂觀，而且也已經有了一些成功的開始。

在一根試管中，或是其它人工的環境中的組織培養，其實已經是老生常談的事情了。著名的 Carrel 博士曾經 "以這種方法來養活一株雞胚胎細胞長達超過三十年之久 (這會比這個胚胎所可能長成的母雞的壽命要長上許多)........." 因著這個明顯的理由，這一株細胞就被稱為 Carrel 的 '不死' 株。然而它在 (疏忽之下) 在二次大戰時期死掉了........。 (87) 完整的器官也已經在動物的身體之外，被養活了各種不同長度的時段，而且在試管中培養器官之想必當然爾，也不需要有多大想像力的延伸。

以一種有些不同但是仍然不失恰當的口吻來說，洛克斐洛醫學研究中心的 Philip Siekevitz 博士表示，"假使在我們有生之年當中，能夠通盤地，然而往往會是特定地，來理解人體如何條控其長成的話，我也不會感到訝異。而去理解，就是意味著去影響。" (105)

We can wait in our timeless freezers for many lifetimes, if need be, but a sufficient degree of control may come fairly early. It need not be based on complete theoretical understanding, but can be to a considerable extent empirical. Experiments based on clever guesses have shown, for example, that embryonic skin treated at a certain stage with vitamin A will develop into the kind of epithelial tissue which lines the intestines, whereas in the absence of vitamin A it forms a normal-appearing skin. (87) Another fascinating news item, bearing on the way in which simple environmental changes can affect reproduction and development, concerns chinchilla breeding. Those bred under ordinary incandescent light produce all-male offspring; under a bluish daylight incandescent light, nearly all progeny are female; in natural daylight, there are equal numbers of each! (99) This particular item invites skepticism, but who knows?

So far we have been talking about growing an organ in the laboratory, starting with a scrap of tissue or a single cell, and then grafting it into the body; but this is not the only possibility. Some parts of the body might be regrown *in situ*.

In lower animals, promising beginnings in organ regeneration have been made. Professor Marcus Singer at Cornell University, by manipulating nerve tissue, has caused adult frogs to re-grow amputated limbs, although normally they cannot.

假使需要的話，我們是可以在沒有時限的冰凍櫃中等待好幾個世代，但是一個足夠程度的控制能力可能很快就會來臨。這並不需要去完全地建構在理論上的了解，而且在某一個程度上，是可以純粹靠著實驗而來的。例如以聰慧的臆測所建構的實驗已經顯示出，在某一個時段，用維他命 A 來處理胚胎的皮膚，將可以轉化成內襯腸道的上皮組織，然而如果沒有了維他命 A 的處理，它就會形成看起來正常的皮膚。(87) 還有一則有趣的新聞，報導有關在小絨鼠的培育中，簡單的環境變化如何能夠來影響其繁殖和成長。那些在一般白熱性光源下培育的，所生出的都全是雄性的後代； 而在一個帶點藍色的日光熾熱光源下，所生出的子代則全都是雌性； 但是在自然的日光下，每種性別的數目是相等的！ (99) 這個特別的情形，可能會招致懷疑，但是天曉得到底是怎麼一回事？

　　到目前為止，我們都一直在談論在實驗室中的事，從組織中的一小塊碎片或是一個單一細胞，開始培養出一個器官，然後再將其接植到身體裡面； 然而這並非是唯一的可能性。身體的某些部分是有可能*在其本來的地方*再度生長的。

　　在較低等的動物中，科學界在其器官的再生上，已經有了一些不錯的開始。康乃爾大學的**Marcus Singer**教授藉著對神經組織的操控，已經可以讓被截肢的成年蛙再度長出，雖然在正常狀況下這是不可能的。

As Dr. Singer says, "Obviously, there is some practical interest in the possibility that human beings might some day be able to re-grow tissues and organs which they presently cannot."(98)

Adult humans can regenerate many tissues (although virtually no organs). Skin is one; another, surprisingly and hopefully, is nervous tissue, at least of certain kinds. In the famous case of the boy whose right arm was severed below the shoulder in a freight train accident in 1961 and sewed back on by Dr. Ronald Malt and co-workers at Massachusetts General Hospital, nerve cells grew back in the arm. In the spring of 1963 healing was not complete, but the cells showed growth at the rate of about one inch per month. (81)

Other pioneering experiments include those at New York University in repairing gaps in human nerves. Frozen nerve grafts from dead donors were used; these were also irradiated to minimize the immune reaction. The grafts themselves are said to work for as long as three years, allowing restoration of muscle function and sensation; eventually the grafts die, but meanwhile there is regeneration of new nerve fibers, which gradually replace the graft. (22)

With such sparkling beginnings, with the quickening pace of research, and with the optimistic outlook of experts, it seems not too much to expect that the brain itself will eventually prove amenable to repair, although enough of the original must remain to preserve memory and personality.

如同 Singer 博士所說的，"很明顯地，在有一天人體可以再度長出組織和器官的可能性中，一定會有某些實質上的用處，雖然這在目前暫時是不可能的。" (98)

成年的人體可以再生出許多種的組織 (然而幾乎沒有辦法再生出任何的器官)，皮膚就是其中之一；另外一種既令人驚訝又覺得幸運地，就是神經組織，至少有某些種是可以的。在一個有名的男孩子的案例裡，他於 1961 年的一次載貨火車車禍中，右手臂從肩膀之下全被撕離了，而由麻州綜合醫院的 Ronald Malt 博士以及其同僚將其縫回，之後其神經細胞有長回到手臂中。在 1963 年春天的時候，雖然他還沒有完全地痊癒，但是這些細胞所顯現出來的成長，大約是在每個月一英吋的速率。(81)

其它的先驅性實驗，還包括有紐約大學在人體神經間隙上的修復。其所使用的冰凍的神經接植體，乃是從死亡的捐贈者而來的；它們都經過輻射處理，來使其免疫反應變得最小。據說這些接植體本身都可以作用長達三年，因此可以來恢復肌肉的功能和感覺；雖然最後這些接植體都會死掉，但是在此同一期間內，已經有新的神經纖維再生出來，這會逐漸地取代接植體的。(22)

有了這些亮眼的開端，有了加快的研究腳步，以及專家們樂觀的預測之下，雖然還是要有足夠的原來的大腦殘留下來，以便能保留住記憶和個性，但是預期大腦本身最終將會被證明是容易修復的，看起來似乎不會是太過分的。

What will no doubt happen, then, is that the more urgent repairs will be made while the resuscitee is still unconscious, with new organs or tissues either grown in the lab and implanted or else gradually regenerated in the body. After this has been done he will be alive, and in much better health than just before he died - but he will still be old.

Curing Old Age

In the discussion heretofore, we have been extrapolating from a solid basis of known achievement. But when we claim that old age will be curable, that senile debility in its varied manifestations will be reversible; we might seem to be on shakier ground. After all, there seem to have been no successes whatever so far in extending human life, except statistical successes based on reduction of infant mortality and conquest of disease.

Nevertheless, theory and expert opinion again provide ample reason for optimism. By way of analogy, one might compare the prediction of a family helicopter with the prediction of an interstellar space ship. The helicopter prediction is conservative; helicopters already exist, and not much daring is required to prophesy that they will eventually become safer, cheaper, and more reliable. On the other hand, no star ship has ever been built or even planned. Even so, interstellar travel is in the cards; if necessary, it can be achieved with chemical fuels and known technology, if we have endless patience and a bottomless purse, but in practice we know we can count on new discoveries as well as the polishing of the old. Interstellar travel is entirely possible in principle, and the practical difficulties will without doubt be overcome; just so with biological immortality.

因此未來無疑地將會有越來越多的緊急修復在被修復者仍處於不醒人事下進行，而其中所用的新的器官或是組織不是經過實驗室中的培養後而移植的，就是在人體內逐漸被再生的。在修復完成之後，他不僅會活過來，而且會比他死之前還要健康許多，但是他將仍然是老的。

醫治老化

在此之前的討論中，我們一直都在一個已知成果的堅固基礎上來進行探索。但是當我們宣稱老化將可以被醫治，各種不同型態的疾病退化可以被逆轉時，我們似乎就陷入一個薄冰深淵的境地。畢竟，除了一些基於嬰兒死亡率的降低和疾病的征服所帶來統計上的成就之外，到目前在人類壽命的延長上似乎還沒有任何成功的案例。

然而再度地，理論上和專家的意見上都提出大量的理由來讓我們樂觀。比喻來說，我們可以來比較一架家庭用直昇機的預測和一架星際太空船的預測。直昇機的預測是較保守的，因為直昇機是已經存在的，而且不需要有太多的冒進就可以預期最終它們會變得更安全，更便宜，並且更可靠。另一方面，從來沒有一架星際太空船被造出來過，甚至於連製造計劃多沒有過。但是雖然如此，星際旅行還是在構想之中；假使需要的話，用化學燃料和已知的科技，如果我們有無限的耐心以及很深的口袋的話，這還是可以達成的，況且在實際上我們知道我們還可以仰賴一些新的發現以及舊科技的翻新。因此在理論上，星際旅行是完全可能的，而且無疑地一些實際上的困難將會被克服；而生物上的永生也是如此。

It is conceivable, although far-fetched, that extended life or even permanent life might result from some kind of "youth serum" such as crops up in the news from time to time. In 1963 a Swiss, Dr. Paul Niehans, is reported treating wealthy old patients with a serum made from the cells of stillborn lambs at $13,000 a shot. (20)

There have been many other reports of possible "youth serums," some of which are still being investigated. For example, in 1963 the National Medical Association heard about extraordinary results in reinvigoration of old people by thyroxine, a hormone of the thyroid gland. Every system of the body is said to be favorably affected, including the circulatory system, nervous system, and digestive system. This research seems to be carried on mainly by Dr. Charles A. Brusch, of Cambridge, Massachusetts, and Dr. Murray Israel of the Vascular Research Foundation, New York, who have treated many patients. They insist that there are no harmful side effects, even with relatively massive doses, and that the metabolism needs this extra spark in old age even when the usual tests ("basal metabolism" and "protein-bound iodine") show normal thyroid function. (See, for example, The Detroit Free Press, August 20, 1963.)

Another sensational report is that attributed in September of 1963 to Dr. Robert A. Wilson, gynecologist of the Methodist Hospital of Brooklyn, New York. He is said to claim, after treating hundreds of patients, that two female sex hormones (estrogen and progesterone), when properly augmented and supplemented with special diets, vitamins, minerals, and exercises, can benefit older women immensely.

就像偶而會出現在新聞中的報導一樣，可能會有某一種"青春血清"來達到壽命的延長或甚至是永久的生命，這雖然有點扯，但是卻是可以想像得到的。在 1963 年，瑞士的 Paul Niehans 博士就曾經被報導使用胎死腹中的羊羔的細胞所製造出來的血清，以一針美金 13,000 元的價錢來治療一個富有的老病人。(20)

另外還有許多有關可能的 "青春血清" 的報導，其中有些是還在研究當中的。例如在 1963 年，國家醫學協會就聽到過有關利用甲狀腺素，也就是甲狀腺分泌的荷爾蒙，來再度活化老人的神奇效果。據說對身體中的每一個系統都有不錯的效應，包括血液循環系統，神經系統和消化系統。這個研究好像主要地是由麻州劍橋的 Charles A. Brusch 博士和紐約血管研究基金會的 Murray Israel 博士所進行的，他們已經治療過許多的病人。他們堅持認為甚至在相對大的劑量下，這是不會有任何有害的副作用的，而且就算在一般的檢測中 ("基礎代謝" 和 "蛋白質結合碘") 都顯示出正常的甲狀腺功能下，老年人的新陳代謝中都需要這種額外的火花。(請參考例如 1963 年 8 月 20 日的底特律自由日報。)

於1963年9月針對紐約市布魯克林Methodist醫院的婦產科醫生Robert A. Wilson博士有另一則吸引人的報導。據報導他在治療過好幾百個病人之後曾宣稱，當兩種女性荷爾蒙 (雌激素和黃體素) 被恰當地增強之後，並且輔佐以特殊的飲食，維他命，礦物植和運動，對較老婦女的助益非常地大。

The secondary effects of menopause are eliminated; heart disease and atherosclerosis are reduced; cancer of the breast or genitals becomes unlikely; the skin improves in texture and color; the bones do not tend nearly as much to become porous and brittle.

There is said to be a specific "juvenile hormone" in certain insects, injection of which will keep them young indefinitely. Although nothing of the kind has been found in mammals, conceivably it might be.

Vastly more likely, progress will come slowly on a mixed theoretical and empirical basis. The theory is just beginning tentatively to be laid, since it is based on obscure, small-scale phenomena which have heretofore been almost inaccessible both theoretically and experimentally.

The electron microscope, the digital computer, the formulas of quantum chemistry, and other experimental and theoretical tools now allow studies on the subcellular level, investigations of the inner workings of the life processes. Biochemistry and biophysics are making violent thrusts in all directions (including, I suppose, backwards).

Drs. B. L. Vallee and E. C. Wacker recently wrote, "Molecular biology, suddenly exploding on cellular biology - as did nuclear physics on atomic physics a generation ago has brought far-reaching challenges and hopes for the solution of questions of normal and diseased life processes thought to be experimentally inaccessible a decade ago." (123)

停經後的次級作用都被消除了；心臟病和動脈硬化症罹患率降低了；胸部或是生殖器官的癌症變成不太可能了；皮膚的紋理和色澤改善了；而骨骼不再那麼容易就變成多孔和酥脆。

在某些種類的昆蟲中據說有一種特別的"年輕荷爾蒙"，注射它將可以使它們永遠保持年輕。雖然類似的東西還沒有在哺乳類動物中找到，但是可想像到地這是有可能找到的。

進展更是大有可能地會在一個混合著理論的和實驗的基礎架構上逐漸慢慢地來到。而其理論只是剛開始暫時地被訂定中，因為其乃是架構在一些曖昧的小規模的現象上，而這些現象在此之前幾乎都是理論上或是試驗上都無法接近去碰觸的。

電子顯微鏡，數位電算機，量仔化學方程式，以及其它的實驗和理論上的工具，讓我們現在已經可以來進行次細胞層次的研究，來探究生命過程中的一些內在作用。生物化學和生物物理學目前正在各個方向上製造出劇烈的衝擊 (我認為也包括著反方向上的)。

V. L. Vallee 女博士和 E. C. Wacker 最近寫道，"分子生物學在細胞生物學中的突然地爆開火花，就如同一個世代前核子物理學在原子物理學中一樣，為解開十年前被認為是無法試驗證實的一些正常的和疾病的生命過程上的一些問題，帶來了一些長足的挑戰和希望。" (123)

- 171 -

To get an idea of the ultra-fine work being done, we may quote a fragment from a fairly recent paper by Fernandez-Moran: ". . . the electron microscope can now be used as a powerful tool both for the controlled production and the direct observation of radiation damage in preselected macromolecular regions of hydrated biological systems. Enhanced contrast and high resolution of the order of 6 to 8 A have been achieved in direct studies of the macromolecular organization of virus particles, ribosomes, and of isolated cell constituents." (32) An Angstrom unit, abbreviated A, is a hundred millionth of a centimeter, and there are 2.54 centimeters to the inch! Using such techniques, and others, Dr. Fernandez-Moran has found an "elementary particle," only 80 to 100 A in diameter, which he regards as the ultimate unit of the function of the mitochondria, which are tiny granules or rods located in the cytoplasm or outer portion of cells. (31)

Concerning the specific problem of aging, there have been many suggestive studies. Enough has been learned to encourage Dr. F. M. Sinex, chairman of the biochemistry department of the Boston University School of Medicine, to say, "The present development of biochemistry and biology suggests the question, 'Why do we get old?' may be answered in the foreseeable future. [Certain hypotheses about aging suggest that] preventative therapy . . . is a possibility." (108)

After prevention, the next step is cure. We may also note that even the prevention of further aging in an old brain might be good enough; most of us die with our strictly mental (as opposed to muscular and glandular) faculties still in reasonably good condition. But in all likelihood the aging process in brain and body will prove reversible.

為了要對進行中的超微細工作有一個概念,我們可以從由 Fernandez-Moran 所發表的一篇相當近的論文中引述一個片段:"……… 現在不管是用在控制性的生產上,或是用來直接觀察一些水化合生物系統中一些預先選定的大分子區塊上的輻射傷害,電子顯微鏡已經可以用來當作一種非常強力的工具。在直接研究病毒,核醣體和細胞分離出來成分諸種大分子組織中,對比的提昇以及 6 到 8 A 等級的解析度都已經達到了。" (32) 一個埃 (Angstrom) 單位,簡寫 A,乃是一億分之一公分,而在一英吋中就有 2.54 公分!應用這種以及一些其它的技術,Fernandez-Moran 博士有發現到一種"基本粒子",其直徑僅有 80 到 100 A,他將其視為是粒腺體運作上的最小單元。粒腺體是位在細胞質或是細胞靠外部份中的一些微小的顆粒或是桿狀物。(31)

針對老化這個特定問題,已經有許多假說性的研究。所掌握到的知識已經夠多,因此激發波士頓大學醫學院生化系系主任 F. M. Sinex 博士說,"由目前生物化學和生物學的發展看出,有關'我們為什麼會老呢?'這個問題,在可見的未來可能就可以得到解答。某些有關老化的假說認為預防性的療法……是有可能的。"(108)

在預防之後,下一步就是治療。我們可能也應該注意到,甚至於在一個老人腦部上進行進一步老化的預防就足夠了,大多數的人死時,其完整的心智官能都還是在相當好的狀況(相對於肌肉和腺體的官能來講)。何況腦部和身體老化的進程將極有可能被證實為是可逆轉的。

There are many ideas regarding the cause or causes of biological aging, and a few of these will now be briefly reviewed, with no attempt to be systematic, let alone exhaustive.

One of the major primary or secondary causes of death usually associated with old age is atherosclerosis, often thought of as "hardening of the arteries" and roughly analogous to scaling or rusting of pipes in plumbing. In recent years, much attention has been given to the suspicion that its development is related to intake in the diet of saturated fats leading to the presence of cholesterol in the blood. However, this view does not seem to be held any longer by a majority of scientists. (113)

In fact, it is amusing to note that the unsaturated fats may be the dangerous ones. According to a Dr. Bernard L. Strehler of the National Heart Institute, ". . . the unsaturated fats are particularly liable to cross-reaction and linkage, a fact that makes them extremely useful in the paint and varnish industry but which may be highly detrimental to biological systems over the long run. The gradual accumulation of a layer of varnish over various intracellular structures is an unpleasant prospect. The observation that the rate of accumulation of cardiac lipofuscin [fatty pigment] is higher in the Japanese, who incidentally consume a diet richer in unsaturated fats, is suggestive. " (113)

Another relatively ancient theory is that aging may be the result of somatic mutations brought on by radiation or other cause. That is, changes in the genetic structure of the body cells may occur haphazardly, from time to time, caused by cosmic rays or other natural radiation (or by fallout from nuclear bombs); since the mutations or changes are almost always for the worse, the percentage of defective cells mounts.

有關生物老化的原因或是多重原因已經有許多說法，現在就其中的一些來做簡要的評論，此評論沒有企圖要是系統性的，更沒有要是長篇大論的。

通常和年邁有關而造成的死亡，其主要或是次要的原因之一就是動脈硬化症，此常常被認為就是"動脈的硬化"而且有點類似管線系統中管子的結垢或是生鏽一般。在近幾年裡，許多注意力都集中在懷疑此症的產生是和日常飲食中攝入的飽和脂肪，而導致血液中膽固醇的出現有關。然而，這個觀點看起來似乎沒有被大多數的科學家認可多久。(113)

事實上，很好笑地反而一些非飽和脂肪才可能是一些危險分子。依據美國國家心臟學院 Bernard L. Strehler 博士，"....這些不飽和脂肪是特別容易交互反應並且產生連結，此事實使得它們在塗料和油漆產業中極度有用，但是就長期而言，對生物系統可能會具高度的傷害性。在各種不同細胞內部結構體上日積月累上一層亮光漆，的確是一種不舒服的感受。科學觀察發現心臟脂褐質﹙脂肪色素﹚的累積速度日本人較高，剛好在他們的飲食中有吃進較豐富的不飽和脂肪，此乃是一種端倪。"
(113)

另外一種相對上較老舊的理論乃是老化可能是由輻射或是其他原因所產生的體細胞變異而導致的。這就是說身體細胞的基因結構，會因著宇宙射線或是其他自然輻射線（或是核子彈的落塵）不時隨意地改變，既然突變或是變異幾乎常常都是較壞的，因此瑕疵細胞百分比會上升。

This theory has some attractions, not the least of which is the fact that animals subjected to heavy doses of radiation exhibit symptoms resembling those of accelerated aging. Nevertheless, this theory has been fairly well demolished by Muller (79) and others.

One of the currently respectable theories seems to be that of Dr. Sinex, who thinks aging may be related to changes or breakdowns in irreplaceable molecules of protein in collagen, which is the main organic constituent of connective tissue. Contrary to a popularly held notion, it is not true that all of the material of our bodies is continually replaced and renewed; it is not true of cells and it is not true of molecules. The same brain cells last us throughout life, and in collagen, at least in rats, the same molecules persist throughout life, or enjoy only limited replacement. If these are damaged by chemical, mechanical, or thermal accidents, the road is downhill. (107)

Another idea is that an "autoimmune reaction" occurs with age; roughly, that is, we can't stand ourselves any more. Still another is that the various subsystems of the body from time to time are taxed beyond their ability fully to recuperate, so that each such part is like a ball which on successive bounces doesn't reach quite as high, and eventually doesn't bounce at all. And so on.

The point is that much has been learned, many promising lines of inquiry are being followed, and as Dr. Joseph W. Still has said, "Medical experience has taught us that when we fully understand a chemical event, we are able to manipulate and alter or modify it. For this reason, we can be skeptical about the assumption that "we can't live forever!" (111)

除了至少在事實上動物接受高輻射劑量會呈現出類似加速老化的諸症狀外，這個理論還頗具吸引力。然而這個理論幾乎已經被 Muller (79) 等人顛覆了。

目前被推崇的理論之一似乎就是 Sinex 博士的了，膠原蛋白乃是結締組織中主要的有機組成物質，而他認為老化可能和其中不可替代蛋白質分子的改變或是破壞有關。和一般認知觀念相左的，我們身體中並非所有的物質都會持續地替代和更新，細胞是如此，分子也是如此。同一套腦細胞我們要用一輩子，而膠原蛋白至少在兔子上，相同的分子要用一輩子，或僅會有有限的更換。假使這些物質有被化學，機械或是燒燙性的意外而受損，生命路途就會走下坡。(107)

另外一個觀念就是隨著老化，一種"自身免疫反應"就會發生，粗略地說，也就是我們不再能忍受我們自己。還有一個就是身體的各種次系統，經常地會被操到超過它們的能力來完全復原，因此每一個這種部位就會像一個連續彈跳的球一般無法維持高度，而最終動彈不得。還有其它的觀念。

重點在於，我們已經懂了很多，很多可能的追尋路線也在跟續中，但如Joseph W. Still博士所說，"醫學上的經驗教導我們，當我們能夠完全理解一個化學事件，我們就可以對其進行操控和改變或是修飾。因此，對'我們不可能永遠活著！'的假說，我們不得不存疑。"(111)

Dr. Strehler, a gerontology specialist, although pessimistic about the practical possibility of abolishing aging (in the comparatively near future, presumably), affirms that: "It appears to me that there is no inherent contradiction, no inherent property of cells or of Metazoa [many celled animals, including man] which precludes their organization into perpetually functioning and self-replenishing individuals." (113)

If nothing better were known, brute-force methods of rejuvenation could be employed. That is, brain cells could be grown in the lab, the appropriate information "read in" to them, and then used surgically to replace the senescent cells. Of course, this would have to be done gradually, over a period of time, and even then some tricky philosophical questions might arise; but these questions will be reserved for a later chapter, since they deserve extended treatment.

But again, in all probability brute-force methods will not be required; more elegant methods are nearly certain to be discovered - provided we do not dawdle too much along the way.

In a recent article in the New England Journal of Medicine was the following remark, intended to be humorous: "If age alone were publicized as an eventually mortal cause of degeneration, associations would undoubtedly be organized to seek its abolishment, under huge federal grants." (47) Many a true word is spoken in jest, and precisely this is going to happen, although one cannot be sure whether the funds will be mainly public or private. Dr. Strehler has already made a plea for sponsorship of a long-term program of research into the biological problems of aging, with something like a National Institute of Gerontology at the helm. (113)

雖然老人學專家 Strehler 博士對消彌老化的務實可能性（可能是指在相對較近的未來中）感到悲觀，但他卻肯定說："就我來看細胞或是多細胞動物〔有許多細胞的動物，包括人〕的遺傳特質中，並沒有先天上的衝突會來限制它們組織成一些會永久運作和自我更新的個體。"(113)

假使不知道有其它更好的回春方法，或許可以來硬幹的。那就是，在實驗室中培殖腦細胞，將恰當資訊"讀入"它們，然後用外科手術來更換感應細胞。當然，這一定要在一段時間內緩步就班地進行，就算如此也一定會產生一些詭譎的哲學問題。因為這些問題應該要詳加探討，所以將諸留在後續章節。

但是再說，極有可能根本不需要硬幹，假使我們在一路上沒有過度蹉跎的話，幾乎一定會發現更多較為優雅的方法。

在近期新英格蘭醫學期刊的一篇文章中，有一則如下意在幽默的說辭："假使只有老化被公認為是一種凡人退化的終極原因，無庸置疑地，一些學會一定會組成，在巨大國家經費的注入下，來尋求它的殲滅。"(47) 雖然戲謔，卻說出許多真話，而確實這是會發生的，只是沒有人可以確實知道經費大部分會來自公家或是私人。針對老化上生物學問題的一個長程的研究計劃，Strehler博士已經提出一個贊助的申請，其招牌有點像是國家老人學學院。(113)

There can be no serious question about the trend of events. You and I, as resuscitees, may awaken still old, but before long we will gambol with the spring lambs -- not to mention the young chicks, our wives.

針對所有事件的趨勢，可以說是沒有什麼大問題。你和我這些被復甦者，醒來時可能依舊老態龍鍾，但是不久我們就可和春天的羊羔一起跳躍—更不用去提那些幼齒的，我們的老婆了。

CHAPTER IV

Today's Choices

當今的選擇

CHAPTER IV

Today's Choices

Overall, three great questions concerning the freezer program are being treated: Is it technically sound, so that the frozen will have a good chance of being resuscitated and rejuvenated? Is it feasible on a practical level, raising no insuperable new problems? Is it desirable, both for the individual and for society?

These questions are to some extent inextricable. In fact, they are so intertwined that there is no completely logical order of presentation, since at almost every stage the argument depends not only on what has gone before but also on what is yet to come, and the picture may not come into clearest focus until a second reading. But in order to finish, one must start, and words have to be set down one after another. In a later age, you and I will no doubt learn better methods of communication.

So far we have dealt chiefly with the first question, and in subsequent chapters shall consider mainly the last two. At the present juncture, the reader is asked tentatively to assume more or less affirmative answers to all three questions, and on this basis to consider the immediate opportunities and obligations presented to him as an individual.

What can we do, today, to improve our own chances? How can we give a dying relative his best chance? If a relative dies when we have made no advance preparation and have limited resources, what can be done? How far, in good conscience, must we carry our efforts?

第四章

當今的選擇

總括地說，現在正在探討的有三個有關冰凍計畫上的大問題：在技術上，是否健全到可以讓被冰凍者能有一個很好的被復甦和被回春的機會呢？在現實層面上，是否可行而不會再衍生出一些無法克服的問題呢？這是否是不管個人或是社會所想要的呢？

這些問題在某個程度上，是難以捉摸的。事實上它們是糾葛不清的，所以我們沒有辦法以完全合乎邏輯的次序來敘述它們，因為幾乎在每一個階段中，此爭論都不僅和以前曾經有過什麼，也和未來會有什麼有關，而其真相在下一個數據未得之前，都無法調到其最清晰的焦點。但是為了要能夠有所了解，我們總是要有一個開頭，而且要一個接一個地下定結語。在經過一段時日後，你和我無疑地就會學到更好的溝通方法。

到目前為止，我們主要地乃是在討論第一個問題，而在後續的章節裡主要地將會去思考後面的兩個問題。在目前的節骨眼時，我們要求讀者對所有上述的三個問題，能夠或多或少暫時地假設其答案是正面的，而能夠以個人的立場來在這個基礎上考慮其所呈現出來的立即的機會和責任。

當今我們能夠做什麼來增進我們自我的機會呢？我們如活來給一個垂死的親戚其最好的機會呢？假使一個親戚死在當我們沒有預先的準備以及沒有足夠的資源下，我們又能做什麼呢？在良心的驅策下，我們付出多少我們的心力才夠呢？

The Outer Limits of Optimism

Before going into detail, if we stand back and look at the problem in its broadest outlines, we note that the extreme limits of optimism depend on two questions: (1) Under what circumstances, if any, is the essence or identity of an individual absolutely and forever lost? (2) What limits, if any, will the human race encounter in its technical development, in its ability to manipulate the universe?

As to the first, the answer of the completely dauntless optimist is that in a deterministic universe no information is ever irretrievably lost, since every detail of history is implicit in the present. Thus, just as the past and future positions of the planets can be calculated from present observations, it is always possible, in principle, to find out every minutest detail of a man's life, memories and personality, given a sufficiently fabulous degree of technical competence. (At least, this seems to be true if we ignore the possibility of the universe being finite in extent, and also ignore limits imposed by expanding-universe theories with their "disappearing galaxies".)

Thus the determinist believes it possible, in principle at least, for a sufficiently advanced civilization to infer as much as necessary about any man who ever lived, and either reconstruct him or replicate him, after gathering together either his original atoms or substitutes. As an intermediate case, an Egyptian mummy could be resurrected; as an extreme case, Ug of Ur. (Some of the "philosophical" problems involved will be discussed in a later chapter.)

樂觀的外部限制

在進入細節之前，假使我們站後面一點，而以其最廣的輪廓來觀看這個問題，我們會看到其樂觀的極度限度是會和兩個問題有關：**(1)** 假使有的話，在什麼狀況下一個人的特質或是身分，會絕對地而且永遠地喪失呢？**(2)** 假使有的話，人類在其科技的發展中，和在其操控宇宙的能力上，會碰到什麼樣的限制呢？

針對第一個問題，對一個徹頭徹尾無怨無悔的樂觀者而言，其答案乃是，在一個確定性的宇宙中，既然歷史上的每一個細節都隱含在當下，所以沒有一筆資訊會一去不回地消失。因此，就如同星球過去和未來的位置，都可以由現在的觀測來計算出來一樣，原理上只要有一個足夠完美程度的科技能力，要來找出一個人生命中的每一個微小的細節、記憶和個性也永遠是可能的。(至少，假使我們將宇宙在其廣度上乃是有限的可能性忽略，同時也將宇宙膨脹理論中所加諸的限制以及它們的一些"消失中的銀河系" 忽略的話，這看起來似乎是真實的。)

因此至少在理論上，這些確定論者相信，在將一個人本來的一些原子或是小成分收集在一起之後，一個足夠先進的文明，要來內推出有關一個曾經活過的人的所有需要的資訊，並且要用其來重建他或是複製他是可能的。舉一個普通的例子而言，一個埃及的木乃伊就可以被復活；就一個極端的例子而言，如**Ur** 城市中的 **Ug**王的木乃伊。(其中某些相關的 "哲學上" 的問題將會在後續的一個章節中被討論。)

Of course, the present consensus (but not the unanimous opinion) of physicists is that the universe is not completely deterministic, and that the outlines of events, whether past, present, or future, must in general always remain somewhat blurred, and that individual atoms have no permanent identity.

In this view, there is a theoretical as well as a practical limit to the accuracy with which we can draw inferences about a man and reconstruct or replicate him. But what this limit may in fact be at present unknown, mainly because we do not yet know enough about microbiology.

As to the second question, no one can be sure how much of what is possible in principle will ever become feasible in practice. If a corpse can lie in a freezer for an essentially unlimited time with no strain on its patience, we can hardly set arbitrary limits on future capabilities. But it is better to limit our guesses to the next few centuries, and to those areas where definite technical developments already point the way.

Preserving Samples of Ourselves

There is one obvious way of helping our chances, in case we die in the early years before non-damaging methods are known for full-body freezing. This is to have little snippets of ourselves surgically removed, while we are in good health, and stored at low temperatures with the benefits of protective chemical infusions. These better-preserved samples can be enlarged in culture by the future technicians for use in repairing our damaged bodies.

當然目前物理學家間的共識 (但不一定是一致的意見)
乃是，宇宙並非完全地確定的，而事件的輪廓，不管是過去的，
現在的或是未來的，一般都必須永遠保持著些許的模糊，而且
單一的原子並沒有永恆的特性。

在此觀點之下，在獲得有關一個人的資訊的內部推論中，
以及在重建或是複製這個人上，其精確度就有理論上以及現實
上的限制。但是當今無法知道事實上這個限制到底是什麼，主
要地乃是因為我們還沒有充分地理解微生物學。

至於第二個問題，沒有一個人能夠確確地知道在原理上是
可能的事情之中，有多少將會變成是在事實上可行的。假使有
一具屍體，可以在一個冰凍櫃中躺上一段實質上是沒有限制的
時間，而且在其忍受的過程中沒有產生任何的變異，這樣我們
就幾乎沒有辦法來對未來的可能，任意地設下一個限制。然
而，我們最好還是將我們對未來的臆測限制在未來的幾個世紀
之中，以及在那些已經有確切科技研發所指示的領域裡。

保存我們自己的樣本

萬一，我們是死在無傷害全身冰凍技術被開發出來之早期
的年代時，還是有一種很明顯的方法來增加我們的機會。也就
是當我們還在非常健康的時候，就用手術的方法取出一小小塊
我們自己的樣本，然後利用保護性化學浸漬的優點，將其儲存
在低溫之中。未來的科技人員就可以將此保存良好的樣本，用
組織培養的方法將其變大，然後用其來修復我們損壞的身體。

In the last chapter something was said about growing cultures, if necessary, from the frozen body itself, and this will doubtless be possible, since a certain percentage of the cells are likely to be in reasonably good condition. At the same time, a margin of safety will be added if samplings of the healthy body are frozen separately ahead of time.

In future eras it will certainly be possible to develop any needed tissues or organs from a germ cell, and it should soon become customary for all adults to make deposits of these cells in cold storage banks. Such banks already exist for the male (sperm) cells, and according to Professor Muller a relatively small amount of research might make a similar procedure possible for women. (77)

Advanced biological art should in fact be able to generate any kind of tissue or organ from a somatic cell; a single scrap of skin might suffice. On the other hand, it is conceivable that at a certain stage in history it might be helpful to have samples of many kinds of tissue from many organs of the body.

It might also be desirable to take tiny samples from many regions of the brain, of course recording the location of the source as accurately as possible. As mentioned earlier, a memory trace is thought to be multiply duplicated in various regions of the brain, so that each of many memories can be both left in the brain and stored in a separate sample vault. Whether a significant number of memories can be protected in this way is an open question.

The procedure seems harmless, since in general tiny specimens taken from various regions of the brain apparently leave it undamaged.

在前面的章節中已經討論過一些有關培養組織的事情，假使不得已的話，此樣本也可以來自被冰凍身體的本身。因為其中有某一比率的細胞，還是處於相當美好的狀況，所以無庸置疑地，這一定是可能的。在這裡只是要指出，假使健康身體的樣本可以事先被分別地冰凍好，那麼就可以有多加一層的安全保障。

在未來的世紀裡，從生殖細胞中來培養出任何所需要的組織或是器官，當然是極有可能的，因而所有的成人將其這類細胞儲存在冰凍儲存庫中，應該很快就會變成一種習慣。其實已經有這類的儲存庫存在，來儲存男性的 (精子) 細胞，而依據 Muller 教授，有一個相對較少量的研究，也有可能會來研發出類似的方法來供婦女們使用。(77)

事實上，先進的生物科技應該有可能光從一個體細胞中，就可以生產出任何種類的組織；而且可能只要刮一下皮膚的量就夠了。當然在另一方面，我們可以想像得到，在歷史的某個階段中，從身體越多種器官中去取出越多種不同組織的樣本，一定會是越有幫助的。

去從腦部許多不同的區域中，去取出一些微小的樣本，可能也是值得的，當然其來源的位置，一定要盡可能精確地紀錄下來。如同前面所提及的，科學界認為一個記憶的痕跡，乃是被多重地複製在大腦中不同的區域，因此眾多記憶中的每一個，都不僅可以被遺留在腦中，而且會被儲存在另外一個樣本櫃中。至於用這種方法能否護存足夠數目的記憶，則是一個供開放探討的問題。

因為一般而言，從腦部不同的區域取出的一些微小的樣本，很明顯地是不會造成傷害的，所以這個方法看起來似乎是無害的。

Haldane, for example, referring to the work of Lashley says, ". . . while removal of a large fraction of a rat's cerebral cortex abolished the learnt capacity to traverse a maze, local injury to any small part of this volume had little or no effect. The facts on human cerebral injuries lead to a similar conclusion." (37) (In other words, we are brainier than we need to be, in spite of the daily news headlines.)

Manifestly, this kind of procedure will not soon, if ever, lend itself in full to large-scale application; there are not enough brain surgeons, nor people anxious for brain surgery.

In the near future, as a compromise, perhaps it will become routine during any surgical procedure to take a few extra snips here and there for the bank. In a different but roughly similar way, it is already becoming useful for people with rare blood types to freeze-bank it for use in case of emergency.

Preserving the Information

We normally think of information about the body as being preserved in the body - but this is not the only possibility. It is conceivable that ordinary written records, photographs, tapes, etc. may give future technicians enough clues to fill in missing or damaged areas in the brain of the frozen.

例如 Haldane 引述 Lashley 的研究所說的，"...雖然移除一大部分兔子的腦神經中樞，會讓其喪失已經學會的走出迷宮的能力，而這一區塊中任何微小部分的局部性損傷，則僅會有微小的甚或是完全沒有影響。人類腦部損傷的一些事實，也都指向類似這一個的結論。"(37)(換句話說，不管報紙頭條新聞是怎麼報導的，我們的大腦總是比實際上所需要的還多。)

打開天窗說亮一點，這種方法就算已經有了的話，也不會是很快地就能夠讓其全面性大規模地應用；因為現在不僅沒有足夠的腦外科醫生，也沒有人那麼迫切地要接受這類的腦部手術。

在較近的未來，當再進行任何一種手術的時候，順便額外地到處剪下一些樣本來放在儲存庫裡，或許將可能會變成一種例行公事，而來當做一個妥協的手段。現有一種不同但是大略上是類似的情形，就是將血液冰凍入庫，以應緊急狀況之需，這對具有稀有血型的人，已經逐漸變成非常有用。

保存資訊

我們一般都會將有關身體的資訊，認為是一定要保存在身體裡面─但是這並非是唯一的可能性。想像可及地，普通的書寫紀錄、照片、磁帶等等，可能可以提供未來的科技人員足夠的線索給，來在被冰凍者的腦中，填補遺失的或是受損的區域。

The time will certainly come when the brain's method of coding memories is thoroughly understood, and messages can be "read" directly from nervous tissue, and also "read" into it. It is not likely that the relation will be a simple one, nor will it necessarily even be exactly the same for every brain; nevertheless, by knowing that the frozen had a certain item of information, it may be possible to infer helpful conclusions about the character of certain regions in his brain and its cells and molecules.

Similarly, a mass of detailed information about what he did may allow advanced physiological psychologists to deduce important conclusions about what he was, once more providing opportunity to fill in gaps in brain structure.

It follows that we should all make reasonable efforts to obtain and preserve a substantial body of data concerning what we have seen, heard, felt, thought, said, written, and done in the course of our lives. These should probably include a battery of psychological tests. Encephalograms might also be useful.

Like anything else, this notion can be carried too far. Pushing this kind of reasoning to the extreme, one might say that one need only preserve a single cell of his body, for its genetic content; from this he could be regrown, and the original personality and memories, at least in coarse outline, implanted from the records. But this sort of connection is both too difficult and too tenuous and unsatisfying for most people.

大腦對記憶編碼方法的完全理解,有朝一日是一定會來臨的,而屆時,一些訊息就可以從神經組織中直接地被"讀取"出來,而且也可以"寫入"其中。其間的關聯性絕對不可能會是單純的,甚至於對每一個頭腦而言,也不需要是全然地相同;不管如何,只要知道冰凍者所擁有的某一項資訊,我們就有可能來推斷出有關其腦中某些區域的特徵,以及其細胞和分子的一些有用的結論。

　　同樣地,有關他曾經做過什麼的這一大堆細鎖的資訊,先進的生理心理學家可能就可以從中歸納出有關他曾是何種人的一些重要的結論,這就可以再度地提供更多的機會,來填補其腦部結構中的一些空缺。

　　因此在我們人生的過程中,我們所有的人都應該付諸合理的心力,來獲得和保存有關我們曾經看過、聽過、感受過、思想過、說過和寫過的一個主要的資料體。其中可能還應該包括有一系列心理學的試驗結果。有一些腦部攝影圖可能也是會非常地有幫助的。

　　和其他情形一樣,這個概念也可能會被過度地宣染。如果把這種推理方法推到超過限度的話,有人可能就會認為,只要我們保存人體中的一個單一細胞,來獲取其中的基因內容就夠了;因為由這個就可以再度地培養出一個人,而其本來的個性和記憶,至少在粗略的輪廓上,是可藉著那些紀錄的再度植入而獲得。然而這樣的聯想不僅是太困難,而且是太薄弱了,對大多數的人而言,這是無法令人滿意的。

Yet we can be sure that before long "record mania" will be added to our list of tics, and swindlers will peddle all kinds of bizarre recording devices and services. No advance is without its price.

Organization and Organizations

What practical steps can one take to ensure that he will be frozen at death? A number of obvious courses suggest themselves.

One of the simplest steps is to specify in your will that you insist on being frozen. (A number of people have already done so, as of this writing, I am told, including persons in Michigan, District of Columbia, New York, New Jersey, California, and Japan.) To make sure this demand is effective, of course, a number of precautions should be observed.

First, the will should certainly be drawn with competent legal counsel. Second, the details should be made as explicit as possible, and therefore the will should be periodically updated. Third, promise of cooperation should be obtained from your expected surviving next of kin, preferably in writing. Fourth, you should choose an executor both sympathetic to your desire and capable of vigorous and decisive action, not necessarily a close relative. Fifth, you should provide funds for the purpose, possibly in the form of direct or indirect proceeds of a special insurance policy.

況且我們可以確信的，在我們的怪胎名單中，不久一定會出現一些"紀錄狂"，而且會出現一些江湖騙子，開始販賣各式各樣新奇古怪的紀錄器具和服務。當然沒有任何的進步是可以不需要付出代價的。

除了組織還是組織

一個人要採取什麼實質的步驟，才能確定自己在死亡的時候，可以被冰凍呢？在一些淺顯的課程中，有將此表露出來。

其中有一個最簡易的方法，就是在你的遺囑之中，堅持自己要接受冰凍。(當我在撰寫此書之際，有人告訴我，已經有一些人已經如此做了，其中包括有密西根、哥倫比亞特區、紐約、紐澤西、加州，以及日本來的人。)要確認這樣的要求是有效的，當然我們還是要去遵行一些預防措施。

首先，遺囑當然一定是要由稱職的法律顧問來撰寫的。第二，其中的細節一定要寫得越清楚越好，而且這個遺囑一定要定期地來更新。第三，一定要從你的家族成員之中，預選出下一個還會存活的人，來獲得其協助同意書，而且最好是書面的。第四，你一定要指定一個執行者，他不僅要能夠對你的想法能有同感，同時也要能夠有旺盛和果斷的行動力，而這並不一定要是一個關係親密的親戚。第五，你應該要提供資金來達成這個目的，這可能是以一種直接的，或是間接的特殊保險合約的形式來給付。

Pursuant to the question of money, it is clear that if you are living up to your income or slightly beyond it, as most of us are, you must mend your ways and practice thrift. Your estate, including insurance policies, must provide for any dependents in addition to purchasing freezer accommodations and a trust fund for yourself. A wise and moderate balance must be struck in all things; however, the more money you save, the more you will be able to take with you, and the more influence you will wield in the meantime.

Another obvious step is to obtain the promise of cooperation from your physician in case of death. This is not meant to imply that you should deliver an ultimatum tomorrow that unless he promises to help freeze you, you will change doctors. Most physicians, in the immediate future, will be very skittish on the subject. But you should discuss it with him, make your views clear, make sure he informs himself on the subject, and maintain a judicious pressure. This kind of action, together with other developments, will assure that before too long there will be an ample choice of cooperative physicians. (It is not suggested that physicians are reactionary and ignorant and have to be led by the nose; but they naturally tend to be conservative, and they need to be informed both of specialized technical developments and of patient opinion.)

A whole crop of organizations will undoubtedly sprout in the fairly near future, offering various services, or a whole range of services, in connection with the freezer program. Perhaps some of them will be formed by morticians, or will be adapted from existing mortuary companies.

攸關金錢方面的問題，很明顯地，假使你現在活得和大部分的我們一樣，都是把收入花得精光甚或超過的話，那麼你就應該要修改你的生活方式而來節省一點。你的資產，包括你的各種保險契約，除了一定要提供你的任何一個相關人，有辦法去購買冰凍容納空間之外，還要有一份是留給自己的信託基金。一個明智的以及恰當的財務規劃，一定要能夠打點到所有的事誼；無論如何，如果你能夠存越多錢，你就能夠帶走越多，而在同時你就可以施展較多的影響力。

另外一個明顯的步驟，就是萬一在你死亡的時候，一定要能夠從你的醫生方面，獲得協助的同意書。這當然沒有居心要去影射說，除非他答應幫助冰凍你，否則明天你就要送他一張要更換醫師的最後通牒書。在最近的未來，大部分的醫師對這個問題都將會是非常地戒慎恐懼。但是你還是應該要去和他討論這個，讓他清楚你的觀點，務必使他能夠讓他自己擁有這方面的資訊，而能夠承受做出正確判斷的壓力。用這樣的行動，再加上其他的研發成果，可以讓我們在不會太久之後，就會有較多願意合作醫師的選擇可能。(這當然不是意味著醫生們都是被動的和無知的，而必須要被牽著鼻子走；而是說他們本質上就是傾向於保守的，而且他們必須要被告知的，不僅是特殊的科技進展，也包括病人的看法。)

無疑地，在相當近的未來，一大堆組織將會如雨後春筍般地冒出，來提供和冰凍計劃有關的各種服務，或甚至是整個範圍全套的服務。其中有些可能會是由殯葬業者來組織而成的，或是會由現有的靈骨塔公司改組而成的。

But until commercial organizations are on the scene, people will have to hand together to form their own.

In union there is strength, and existing organizations, for example fraternal societies, could form committees and sub-organizations, possibly somewhat on the order of burial societies, to serve their members. The pool would provide moral, financial, and administrative support. All preparations would be made in advance, making the most and best of local conditions, and on the death or impending death of a member the organization would swing into action.

If in some cases it turns out to be awkward to work within existing organizations, then mutual aid societies can be formed with this specific purpose, the usual legal precautions being observed.

Finally, another way the individual can help the general impetus is to write his life insurance company, inquiring about freezer insurance. Many companies already sell special-purpose policies, for example, with the proceeds ear-marked to pay off a mortgage. In logic, of course, this seems a little silly, since the beneficiary might as well simply have the additional funds, to apply as seems fit; but psychologically the companies find this device useful. Also, it is not clear that the life insurance companies would want, or would be legally able, to have a direct hand in physical freezer facilities. But the point is that there is an immense new market for life insurance, and when this is realized the life insurance companies are sure to exert heavy influence, directly or indirectly.

但是在這些商業性組織出現之前，人們將一定要靠大家合作，來組織屬於自己的組織。

　　團結就是力量，而且例如像是現有的一些交誼性社團，就可以組成一些委員會和次級組織，有可能是有些類似治喪委員會層級的社團，來為其會員服務。這個組群可以提供在道德上、財務上和管理上的幫助。所有的準備都應該事先完成，以便能夠大大地利用週遭的條件，而且當一個會員死亡或是瀕臨死亡之際，這個組織就可以馬上進入行動。

　　假使在某些狀況下，發現要在現有的組織中進行是有點齟齬的話，那麼就可以組成一些針對這個特定目的，而且能夠遵照一般法律預防措施的互助性社團。

　　最後，還有另外一種方法，個人可以幫助其全面性的推動力，就是去寫信給他的壽險公司，詢問有關冰凍計劃保險的事宜。目前有許多保險公司，已經在販賣特殊目的的保險契約，例如說有將理賠指定來付清房貸的。當然在邏輯上，這看起來似乎有點愚蠢，因為受益者大可以單純地將多出來的錢直接花在他自己認為恰當的事情；但是在心理上，這可以讓壽險公司來認識到這種設施的好處。而且，我們也不清楚，保險公司是否會想去，或是其在法律上，可以直接去碰觸冰凍設施的一些實體運作。然而重點在於，對壽險業而言，這是一個很大的新興市場，而當一切成真時，壽險公司一定會直接地或是間接地施展出重量級的影響力。

Emergency and Austerity Freezing

Many circumstances of death, in the near future, will pose a painful and nearly intractable problem for the next of kin. Substantial funds may be lacking; medical cooperation and hospital facilities may be lacking; death may come unexpectedly and the body may not be found immediately. What can be done in such cases, and how much hope do the possibilities afford?

The second question has already been discussed. In the worst cases, most scientists would doubtless characterize the chance of revival as remote or even vanishingly small; but this estimate is based on a feeling and not on a calculation.

The estimate can perhaps be regarded as depending on three factors. First, is the degeneration really irreversible in principle. Second, how nearly will technical feasibility approach theoretical possibility, looking into the indefinite future? Third, how likely is it that historical developments will deny to the frozen the treatment technology could provide them?

It seems to me that at present we cannot make even a reasonable guess about the first two, while the third, based on discussion in later chapters, has a most hopeful answer. If this reasoning is correct, then estimating the chance as "remote" or "vanishingly small" represents nothing more than a vague and generalized pessimism, arising because many scientists are overawed by the apparent difficulties.

緊急和自我設限的冰凍

在最近的未來，許多死亡的情境，對其親朋好友，將會加諸一個不僅痛苦的，而且幾乎是無法應付的問題。可能會是沒有足夠的錢；也可能會是缺乏醫療的協助和醫院的設施；死亡可能會不預期地來臨，而屍體可能沒有辦法被馬上找到。在這些情況下，我們能做什麼，而且其中的可能性，又能夠給予我們多大的期望呢？

有關第二個問題，我們已經討論過了。在最壞的情形下，大部分的科學家無疑地，一定會把其復甦的機率認定為是渺茫的，甚或是不可見地小；但是這樣的評估，乃是基於其感覺，而不是基於其科學的計量。

這個估計可能可以被認為是和三個因素有關。第一，在原理上，敗壞是否真的是不可逆的。第二，前瞻不定期限的未來，技術上的可行性將會多接近理論上的可能性呢？第三，在歷史的進展中，會去剝奪冰凍者，去接受可以提供他們的治癒技術的可能性有多大？

就我看來，在目前我們似乎還是沒有辦法針對前面的兩個問題，做一個合理的臆測，而第三個問題，依據後面章節的討論，則會有一個最有希望的答案。假使這個推理是正確的話，那麼將機率估計為 "渺茫的" 或是 "不可見地小" 所代表的，只不過是一個模糊的和以偏概全的悲觀主義，此乃是由於許多科學家，被一些表面上的困難給過度驚嚇到了。

Even so, in the immediate future it would take an unusually strong and resourceful person, with nerves of steel, to undertake freezing single-handed. If a mutual aid society, or even a coherent family, can work together, however, probably something can be done, and a few practical suggestions will now be offered.

It is understood that these suggestions do not constitute medical advice, carry no guarantee of any kind, and are not even claimed to represent a consensus of current opinion. They represent only the author's impressions, as of this writing, for whatever they may be worth. The reader is expected to seek other opinions, as recent and as authoritative as may be.

First, whoever is present at time of death, or soon after, should probably try to reduce the rate of deterioration by applying artificial respiration and external heart massage. (Tubes are available for mouth-to-mouth artificial respiration without actual contact; sources of information on these techniques can be obtained from physicians, druggists, and libraries.)

A physician should be called as quickly as possible to certify death. Then cooling and freezing should be accomplished by the best available means. Ice might be used at first if nothing else is at hand, or the body might be placed in a cold room in winter. Dry ice might be used next, being readily available in all cities during business hours, at a price currently of around 6 cents a pound or less. The body might be packed in dry ice chips, with blankets to maintain contact and keep out the heat; or faster cooling might be accomplished by using one of various slushes of liquid chemicals mixed with dry ice.

儘管如此，在即刻的未來，一定要出現一位意志非常堅定，而且資源豐富的人，能夠用其鋼鐵般的膽量，來獨撐冰凍計劃的重責大任方可。然而，假使有一個互助的社團，甚或是一個意見一致的家庭，可以團結合作的話，或許還可以做一些事情，而針對此，我們現在要來提供一些實際的建議。

　　大家要先了解，這些建議並不代表著一種醫療意見，其中不包含著任何形式的保證，而且甚至於沒有宣稱是代表著目前各種意見的一個共識。它們僅是代表著作者目前在撰寫此書時的一些感受，不管它們的份量有多少。讀者應該還要去參尋其它的意見，最好是越新的而且是越具權威的越好。

　　首先，在死亡的時候或是剛死不久，不管是哪一個人在場，可能都應該要藉由人工呼吸和心臟外部按摩，來降低其敗壞的速率。(現在已經有不需要實際接觸的口對口人工呼吸用的管子了；有關這方面的資訊來源，可以從醫師、藥劑師和圖書館中來獲取。)

　　接著要盡快找一個醫師，來確認其死亡。然後要用既有最好的方法來完成冷卻和冰凍。假使手邊沒有其它的東西，先用冰塊也可以，或是在冬天時，也可將身體先放在一個冷房裡。之後，可以使用乾冰，目前每一個城市在營業時段時都買得到，現在其價錢大概是每磅六分美金或是更便宜。可以將身體用乾冰的碎片包覆起來，外加一些毯子來維持乾冰的接觸以及隔絕掉熱量；或是可以用各種不同液態化學物的漿體和乾冰混合，來完成較快速的冷卻。

A few words of caution should be inserted. Communicable diseases, of course, require special precautions. Water should not be allowed to get into the body cavities. Dry ice should be handled gingerly, or with gloves; and while carbon dioxide is not poisonous, if too much is used in too confined a space a lack of oxygen may result. If the body is not discovered until it has begun to stiffen, the artificial respiration and heart massage are probably useless, since the blood vessels are clogged, and this part of the procedure would be omitted.

The problem of where to store the body is one the individual, family, or mutual aid society will have to solve. The question of a container and its cost and servicing will be touched upon in Chapter VII.

Freezing with Medical Cooperation

If medical help and hospital facilities can be obtained, the outlook is much brighter. Various possibilities have been alluded to in Chapter III, particularly that of perfusing the whole body with glycerol solution, along with supportive measures, before freezing the body with liquid nitrogen; this may afford the best chance at present to minimize - although by no means to eliminate - injury.

With a cooperative physician and careful advance preparation, obviously the odds will improve immensely. If the physician hesitates to work on the body himself, he might at least be willing to supervise preparation, have himself or an associate available in the hospital for a very quick finding of death, and train a mortician to do the actual work after death.

在此應該要插入幾句警告的話。針對傳染性疾病，當然需要有特別的預警措施。不可以讓水進入體腔內。使用乾冰時，要非常謹慎，或是要戴手套；而且雖然二氧化碳是沒有毒性的，但是如果在一個太密閉的空間用太多的話，就會導致缺氧。假使身體在已經開始僵硬之後才被發現，人工呼吸和心臟按摩可能就沒有用了，因為此刻血管已經堵塞，所以這部分的程序應該可以被省掉。

至於有關要在哪裡儲存身體，乃是個人、家庭、或是互助社團，必須要去解決的一個問題。關於容器和其成本以及服務的問題，我們將於第七章中有所著墨。

醫學協助下的冰凍

假使可以獲得醫療上的協助和醫院的設施的話，那麼前景就會比較光亮。在第三章中，我們已經有間接地提到各種不同的可能性，尤其是在用液態氮冰凍身體之前，如何用甘油溶液浸漬全身，以及用一些輔助性的方法的事情；這可能可以在目前提供最好的機會，來使損傷變得最小，雖然絕對不是可以完全消除它。

有了一個醫師的協助以及預先細心的準備，很明顯地，其勝算將一定會大大地增加。假使醫生不太願意自己親身去處理屍體的話，他至少可能會願意來監督其準備動作，可以在醫院裡找到他或是他的一個助手，來對死亡作一個快速的鑑定，而且它可以訓練一個殯葬人員，來在死亡之後進行真正的工作。

The mortician, of course, would also have to be quickly available - how quickly, would depend on the methods used; again refer to Chapter III. The state of the art is constantly improving, and new and better methods may be known by the time this book is in print.

Physicians may often be reluctant to cooperate in freezing for several reasons - generalized fear of criticism, fear that they lack competence in the techniques, and fear that all their dying patients will demand freezing. None the less, some physicians are willing to try desperate or experimental measures in otherwise hopeless circumstances, and it is possible to put the case in this light. That is, if the patient is in a hospital and known to be near death, the physician might be persuaded to give medical help on the basis of treatment.

We recall that "suspended animation" is ordinarily taken to mean freezing without damage, so the person is regarded as still alive, and capable of being revived at any time without waiting for new developments; this technique is not yet perfected. But we also recall the experiments with whole-body perfusion of rats, and other evidence suggesting that people could be cooled by passing cold glycerol solution through their circulatory systems, and then stored at low temperatures in relatively good condition, although at present we have no means safely to thaw them and remove the protective agents. It is possible, although not certain, that the greater part of the damage occurs in thawing and not in freezing; hence these patients, after freezing, need not definitely be considered dead, and their condition could be called "suspended animation".

當然這個殯葬人員也要很快就可以到達，至於需要多快，則會和所使用的方法有關；請再度參讀第三章。技藝的水準乃是隨時在進步的，而在此書付梓之前，搞不好又會發現一些較新較好的方法。

　　醫生因為有許多原因，通常可能會不太願意來協助冰凍─例如，一般性地對批評的懼怕，害怕他們缺乏這類技術上的能力，並且害怕他們所有的病人都將會要求要冰凍。但是無論如何，總是會有一些醫生在不然就完全無望的情境下，會願意來嚐試一些死馬當活馬醫的或是試驗性的方法。而這種情形或許可以用這樣來說明，那就是假使在一家醫院的病人已經知道是瀕臨死亡，那麼醫生就有可能被說服，以治療的名目來給予醫療上的協助。

　　我們回顧所謂的"活體休眠"，一般乃是用來指沒有傷害的冰凍，因此這個人是被認為還是活著的，而且不需要等待新的研發，就可以來被復活；這種技術目前是尚未成熟完美的。但是我們也回顧一下有關兔子全身浸漬的實驗以及其它的證據，這些都指出，我們可以藉著讓冰冷的甘油溶液，流通過人體的血液循環系統，而來冷卻他們，然後，再以相對好的狀態儲存在低溫之下，儘管雖然在目前，我們還沒有完全安全的方法來將他們解凍，以及移除那些保護製劑。雖然我們還不能確定大部分傷害的來源，但是它們可能都是發生在解凍的過程中，而不是在冰凍過程中；因此這些病人在冰凍之後，不需要硬生生地將其認定為死亡，而且他們所處的狀態堪被稱為"活體休眠"。

Thus some courageous physicians, if persuaded by patient and family, might agree to freeze the subject before natural death, with all the advantages of deliberate preparation and a body in better condition; the purpose would be to reduce metabolism and preserve life while a cure was sought. No death certificate would be issued, and the freezee would remain a patient and not a corpse, with various legal and practical advantages - and also, of course, some disadvantages.

A variation of this idea might be to have one physician certify death, after the patient expires, and a second physician immediately treat the body to prepare it for freezing, later certifying that in his opinion the patient may not be dead. The major biological advantage, of treating a fully living body, would be lost, to be sure, but this might be necessary to induce the physician to cooperate. The death certificate would help protect the second physician, while his doubt about the patient's death might be translated into legal life, although the earliest cases would involve protracted litigation.

Individual Responsibility: Dying Children

Many Americans and Europeans, as well as others, will very soon be called upon to make life or death decisions. Perhaps some are facing such decisions this very day, as you read these words.

Let us consider first the most tender and least exculpable example; the impending death of a child.

因此，假使能被病人和其家屬說服的話，會有某些勇敢的醫師，就有可能會同意在一個人進入自然死亡之前，來對其進行冰凍，以便能爭取到能夠來細心準備和身體狀況較好的優勢；其目的乃是要去在尋求治癒療法的同時，能夠降低其新陳代謝以及保存其生命。這將不會需要去簽發出死亡證明書，而且被冰凍者還會被維持是一個病人，而不是一具屍體。其中一定會有各種法律上和實質上的優點，然而同時當然也會有一些缺點。

這個概念的另一種變化方式，乃是在病人死亡之後，由一個醫師來證明其死亡，而由第二個醫師馬上來處理其身體，準備來進行冰凍，之後再依據他的意見，證明此病人可能還沒有死亡。但是我們要確切地知道，如此一來，能夠治療一個是完全活的身體的這種主要生物學上的優勢，將會喪失，然而這可能是要能夠誘發醫師願意來協助所必要的。雖然最早期的一些這種案例，可能會牽扯到冗長耗時的法律訴訟，但是此死亡證明書卻可以用來幫助保護第二個醫生，而他對病人死亡的懷疑，可能可以被解釋成一種法律上的生命。

個人的責任：垂死的小孩

許多的美國人、歐洲人以及其他地方的人，不久都將會被要求來作一個生或死的抉擇。可能當你在讀這本書時，就在此時此刻就有一些人正在面臨這種抉擇。

讓我們首先來思考這個最細膩而且最難辯護的例子；也就是一個瀕臨死亡的小孩。

Every year in the United States, over 150,000 children under nineteen are taken by death, often signaled well in advance as a result of incurable disease. In 1959, cancer alone claimed over 10,000. (124)

Until now, parents could only seek religious comfort, or compose their minds according to their resources. Now it is better, and of course worse. Better, because there is hope. Worse, because hope implies also trouble, turmoil, and the possibility of failure.

If an adult is dying, it can be argued that he should be allowed to make his own decision about freezing; and if he is of advanced age, the rationalization of a "full life already lived" can be used to justify inaction. But in the case of a dying child, the parent cannot easily find shelter from his responsibility.

I realize very well the cruelty of adding to the burden of grief a further torment of difficult decision and a potential load of guilt. Many people will have no clear idea of what is right. On the one hand, it will seem to them, if they freeze the child their hopes may prove unfounded and they will have engaged in gruesome, bootless, agonizing and expensive sacrilege. On the other hand, they may find it hard to forgive themselves if they bury the child and the freezer program nevertheless gains acceptance. It is my view, of course, that the freezer program will become general and will prove successful, and that the price in money and temporary emotional upset is not too high.

每年在美國有超過150,000個小於十九歲的小孩死亡，通常事先都有明顯的訊號，指出其乃是由無藥可醫的疾病所導致的。在1959年，光是癌症就要上10,000條小命。(124)

直到現在，父母親也只能尋求宗教上的安慰，或是藉著他們所擁有的資源來使其心靈恢復平靜。但是現在已經變得較好的，而當然同時也變得較壞了。較好的是因為有希望了。而較壞的是此希望也隱含著麻煩、困擾、以及失敗的可能性。

假使一個成人快死時，我們可以辯論說應該要容許他來作他有關冰凍的抉擇；而且假使他已屆高齡的話，可以用一個"已經活過整個人生"的合理化論調，來決定不進行冰凍的恰當性。然而，在一個瀕臨死亡小孩的情況中，父母親不可能很容易地就可以找到他的責任的避護所。

我可以深深地體會到，在傷痛的重擔上，再進一步地加上面對困難抉擇的折磨，以及可能有的罪惡感負擔的殘忍。有許多人將會搞不清楚什麼才是對的。在一方面，他們似乎會覺得，假使他們將小孩冰凍，他們的希望可能會被證明為是沒有根據的，而他們將會被牽扯入一個令人毛骨悚然的、沒有用的、痛苦難忍的、而且又是昂貴的一個悖理逆天的行為。但在另一方面，如果他們把小孩埋了，然而冰凍計劃畢竟卻是逐漸被接受的，他們將會覺得難以原諒自己。當然這是我自己的觀點，冰凍計劃將會變得普遍而且將會被證實是成功的，而金錢上和短暫情感上的不舒服的代價，不會算是太高的。

The decision will necessarily be on an individual basis. Entering into it will be such considerations as estimate of the chances, advice by physicians and clergymen, the status of the freezer program in general, and the financial and emotional situation of the family.

Assuming parents find the strength and resources to freeze a child, and eventually see him safely in a permanent Dormantory, they will then have time to ponder some very disturbing questions. When will I see my child again? If I die at an advanced age, will my revival be more difficult than his, and hence later, and will he therefore be older and wiser than I when I awaken? Will the relation of parent and child be effectively reversed? Or will I be frozen by more advanced methods, and therefore revived first, as a physically young adult, and him later, still as a child?

One can only assume that society will gradually evolve a standard operating procedure for dealing with such matters wisely, taking into account both the wishes of the individuals involved and the welfare of the community.

Husbands and Wives, Aged Parents and Grandparents

If your husband or wife is dying, the problem is in some respects different. If the dying spouse wants to be frozen, clearly you should comply, even at substantial financial sacrifice. (One hopes there may eventually be tax relief or subsidy for the families of the early frozen, who have not had the opportunity to buy freezer insurance policies.)

這個抉擇一定要是由個人來決定的。要進入次計劃，一定要有一些考量，如勝算的評估、醫師和牧師的諮詢、整體冰凍計劃的現況、以及家族財務上和感性上的狀況。

假設說，父母親找到了來冰凍一個小孩的勇氣和資源，而最後也看到他能夠安全地被放置在一個永久性的休眠所，如此一來，他們就會有時間來思考一些較費心思的問題了。如什麼時候我才可以再見到我的小孩呢？假使我死在高齡，我的復活將會比他的更為困難嗎？而在此之後，當我被喚醒時，他是否將因而會比我老，並且比我有智慧呢？雙親和小孩的關係是否會因而實質地被顛覆呢？或是，我是否會用較先進的方法冰凍，因此會先被復活成一個身體年輕的成人，而他會較晚被復活，而仍然是一個小孩呢？

我們只能假設，整個社會將會逐漸地演化出一套標準的作業程序，來智慧地應付這些事宜，同時不僅會考慮到相關個人的意願，也會考慮到群體的福祉。

夫妻，年邁雙親和祖父母

假使你的老公或是老婆快死了，那麼這問題在某些方面則是不同的。假使臨死的伴侶想要被冰凍，很清楚地，你就應該要遵從，甚至於在財務上要付出很大的犧牲。(我們希望最終對先期被冰凍的家庭，可能會有減稅或是補助性的措施，因為他們根本沒有機會來購買到冰凍計劃的保險契約。)

If your husband or wife is mentally competent but opposes freezing, a difficult moral problem arises. The easy way out is compliance and burial, but you will have to live with your conscience a long time. The key consideration, it seems to me, is that burial is final, whereas freezing commits one to nothing except a second chance; there is always time to bow out, if one should insist. You can change your mind after freezing, but not after burial.

In the case of an aged parent or grandparent, lacking in vigor and perhaps limited in understanding, there may again be an unwelcome responsibility. Should his decision prevail, or your judgment? Many circumstances will enter, as in the case of children. In addition, the responsibility may be split among several children who may not concur, and one must decide for himself how much effort conscience demands. But the rationalization of "a full life already lived" will not hold water: in the long view, eighty or ninety years is not a full life, but only a beginning.

Even before custom gives sanction, I believe a sufficient number of people will prefer beginnings to endings.

假使你的老公或是老婆是精神上正常的，但是卻是反對冰凍的，那就會產生出棘手的道德問題。簡單避開此問題的方法，就是去遵從其意然後來進行埋葬，但是如此，你將必須要去面對自己的良心許久。依我的觀點看來，思考的關鍵乃在於，埋葬就是一切的終結，然而冰凍，除了給人一個二度機會外，卻是無傷大雅的；假使你真的想要的話，隨時都可以退出。冰凍後，你還可以改變主意，但是埋葬之後，則不然。

針對一個老邁的父母親或是祖父母，由於他們已經缺乏活力，而且可能理解力也有限，因此這可能又會是一個吃力不討好的責任。是應該遵照他們的抉擇呢？還是依據你自己的判斷呢？就如同在小孩子的案例一樣，這會有許多狀況出現。況且，此責任可能會由幾個兒女來分擔，其中有人可能會不同意，因此，一個人可能要為自己裁決要盡心力到何種田地，才符合良心的要求。但是他們"已經活過整個人生"的合理化論調，將不再有用：以長遠的眼光來看，八十歲或是九十歲，將不再是一整個人生，而可能僅是一個開端。

甚至於在傳統習俗能夠接納之前，我相信還是會有夠多的人，將會是較喜歡這是開端，而不是終結。

CHAPTER V

Freezers and Religion

冰凍人和宗教

CHAPTER V

Freezers and Religion

At first thought, one might expect that many religious people will be repelled by the freezer program, refusing to share in it and even denouncing it as immoral. After all, there are several obvious ways in which the program may seem incompatible with religion, if one thinks hastily and superficially.

First, the idea that death is not absolute and final, but a matter of degree and reversible, seems to do violence to the notion of "soul," to the duality of body and spirit which plays an important part in most religions. Might it not be claimed that a freezee, after revival, would be a soulless monster or zombie? Or that to revive a corpse, and thereby recall a soul from its resting place, would be an act of blasphemy?

Second, there is implicit in the freezer program the view that modern man is not the acme of development, but represents only a rung on the evolutionary ladder; that we not only evolved from lower forms of life, but will continue to ascend, through manifold biological and bioengineering techniques, both racially and individually, changing profoundly in both outward and inward nature. Does this not put a severe strain on the idea that man was created in God's image? In particular, can a Christian accept the notion that Jesus, in His human form, did not represent the pinnacle of development?

第五章

冰凍人和宗教

剛一思及此時，我們可能預期會許多有宗教信仰的人，將會排斥冰凍人的計劃，會拒絕參與其中，並且會斥責其為不道德的。畢竟，假使你草率地而且膚淺地來想的話，在此計劃中是有一些明顯的做法，看起來似乎是和宗教不相容的。

第一，死亡不是絕對的和最終的，而是一個程度性的，並且是可逆的概念，對"靈魂"，對肉體和心靈二位一體的這個觀念，似乎會產生極大的震撼，因為這在大多數的宗教裡，都扮演著一個重要的角色。一個冰凍者在被復活之後，難道不會被認為是一個沒有靈魂的怪物或是僵屍嗎？或是說，將一具屍體復活，然後再從其永久安息之地召回一條靈魂，會不會是一種褻瀆的行為呢？

第二，在冰凍人計劃中隱含著現代人還不是發展的顛峰，而僅是代表著進化階梯上的一個台階的觀點；我們不僅是從較低級型式的生命演化而來的，而且是將會通過多重的生物和生物工程技術，讓個體以及整個種族繼續地向前演化，不僅是在外觀上，同時是在內在的本質上，都將會有極深遠的改變。這對人乃是依著神的形象而被創造的信念，難道不會產生出一個嚴重的變化嗎？尤其是，一個基督徒能否夠接受當耶穌在其化身為人時，所代表的並非是發展的頂點的意念呢？

Third, some churchmen will see looming larger the specter of creeping secularism. With unlimited physical life in prospect, will the flocks forget about spiritual immortality? Will they turn en masse to materialism? Will they worship only the Golden Calf?

Several subsidiary and related questions also present themselves.

Forbidding as these questions may appear, 1 believe they will evaporate rather quickly, leaving behind only a few patches of fog which will continue to swirl for a long time.

Revival of the Dead: Not a New Problem

Hundreds of people have already been resurrected from the dead, with no fuss or question as to the abode of the soul during and after death. These were the victims of drowning, asphyxiation, heart failure, and the like, who suffered clinical death but were revived by the use of artificial respiration, heart massage, chemical stimulation, electrical stimulation, and other methods of modern medicine. An especially interesting case is that of Roger Arnsten, a Norwegian boy who drowned in 1962 and was dead for about 2.5 hours, including an estimated twenty-two minutes under water.

Roger, five, fell into an icy river on a cold winter's day. After drowning, his body temperature continued to fall, probably getting below 75F, and of course this hypothermia prevented swift deterioration of his brain.

第三，有一些教會人士會看到潛伏的世俗主義妖魔，將籠罩著越來越大的陰影。有了無限長肉體生命的前景，人群大眾是否會忘掉了靈魂的不朽呢？他們是否會全部轉向物質主義呢？他們是否將僅會崇拜金造牛犢呢？

還有許多附屬的和相關的問題也自然會出現。

這些看起來可能都是非常禁忌的問題，但是我相信它們很快就會蒸發消失了，殘留下來的僅會是幾片霧塊，雖然它們還將會盤旋許久。

死者復活：這不是新鮮事

已經有好幾百人被從死亡中救回來了，然而對在其間和其後，靈魂到底跑到哪裡去了，都沒有人大驚小怪或是發出疑問。這些人都是溺水，窒息，心臟病，以及其他類似情形的受害者，他們都曾蒙受醫療臨床上的死亡，但是藉著使用人工呼吸，心臟按摩，化學藥品刺激，電力刺激，以及其它現代醫學的方法，都被再度復活了。其中有個特別有趣的案例，有一個挪威籍的小孩 Roger Arnsten，他在 1962 年時遭到溺斃，死亡了大約 2.5 小時，而其中估計大約有 22 分鐘是在水下的。

五歲的Roger在酷冷冬季的一天，跌入了一條冰冷的河流。在溺水之後，他的體溫持續地下降，可能達到華氏75度以下，當然因此這種失溫現象使他的大腦免於快速的敗壞。

Dr. Tone Dahi Kvittingen applied artificial respiration with a tube down the windpipe, and rhythmic pressure on the chest to force blood circulation. At the hospital, an electrode needle pushed through the chest wall into the heart revealed no beat; but the attempt at resuscitation was continued, including exchange blood transfusions, and about 2 1/2 hours after drowning a natural heartbeat resumed. In the sequel, Roger remained unconscious for about six weeks, and even went temporarily blind, and at times appeared demented, but finally made a nearly complete recovery, with slight impairment of some muscular coordination and peripheral vision. (58)

The point here is that nobody worried about little Roger's soul. Did God, knowing he would be revived, rule that this was not really death and simply leave the soul in the body? Or did He keep the soul in escrow, as it were, and return it to the body at the moment of resuscitation? If the boy did leave his body temporarily, was he conscious or unconscious? No one knows, and no one seems inclined to make an issue of it.

Why, then, should anyone be concerned about the souls of the frozen? The mere length of the hiatus can hardly be critical: in God's view, 300 years is only the blink of an eyelash, and presents no more difficulty than 2 and ½ hours.

Except quantitatively, then, the problem is not new, and the religious communities have already made their decision. They have implicitly recognized that resuscitation, even if heroic measures are employed, is just a means of prolonging life, and that the apparent death was spurious.

Tone Dahi Kvittingen 醫師用一根管子深入其氣管來進行人工呼吸，而且以具節奏性的壓力來壓其胸部，以強迫其血液循環。在醫院裡他用一根電極探針穿過胸腔，進入其心臟來探測，然而並沒有顯示出心跳；但是救復的努力還是繼續，其中包括了換血輸血，而大約在溺斃 2 1/2 小時之後，一個自然的心跳恢復了。接著大約經過六個星期，Roger 還是一直沒有意識，甚至於變成暫時失明，而且有時候會出現精神失常，但是到最後，竟然幾乎完全康復了，僅有某些輕微的肌肉協調和周邊視覺的傷害。(58)

重點乃是沒有人曾經去憂心過小 Roger 的靈魂。上帝難道知道他會被救活，而判定這不是一個真正的死亡，因而乾脆就把他的靈魂留在他的身體裡面嗎？或是祂先將其靈魂以其現況暫時監管起來，而在復甦的時刻，再將其放回他的身體裡面嗎？假使這個男孩確實曾經暫時離開他的身體，那他當時到底是有意識還是無意識呢？這是沒有人知道的，而且也沒有人似乎會想去對此大驚小怪。

那麼為什麼有人會去顧慮被冰凍者的靈魂呢？其間片段的些微長度，幾乎是無關緊要的：以神的觀點來看，300 年的時光其是僅是一眨眼的片刻。而且其間所存在的困難，並不會比兩個半小時間的還要多。

因此，除了在計量的方面外，這個其實不是一個新的問題，而且宗教的社群也已經做了他們的判定。他們已經含蓄地認定，復活術，甚至於要將其應用到較大膽的做法，也僅是一種延長生命的方法而已，而且外觀上的死亡是騙人的。

The Question of God's Intentions

The cry will certainly be raised in far right religious quarters that freezing is "unnatural" and that it was not "intended" for cadavers to be revived. The answers to this should be quite obvious, but we may as well indicate them anyway.

Part of the answer lies in a recent version of a very old joke. A querulous lady objects to astronauts attempting to leave God's green earth for outer space. "It's against the will of God," she says, "for man to try to live in the sky, going to the moon and Mars and such. Why can't those people just stay quietly at home and watch TV, like God intended?"

A somewhat earlier version concerns objections to Henry Ford's Model T. "If God had intended man to go forty miles an hour, He would have provided him with wheels instead of legs."

This attitude is less amusing in the case of certain sects said to oppose the "interference" of physicians in the course of nature, even forbidding the use of silver nitrate in the eyes of the newborn, on the ground that God "intended" the child of a gonorrheal mother to be blinded.

It is exactly man's nature to "go against nature." Beasts live, even though miserably, in "harmony" with nature; but man must strive to improve both himself and his environment.

上帝旨意的疑問

在極右派宗教陣營中，當然會有人拉高嗓門呼叫說冰凍是"違反自然的"，而且老天本來不是有""意圖"要讓屍體被復活的。針對這情形的答案應該是非常明顯的，但是我們倒不如還是把它們表達出來。

在一個老掉牙的笑話裡的一個新近版本中，就可以找到一部份的答案。有一個喜歡發牢騷的女人，她反對太空人試圖要離開上帝的綠色地球，而來飛向外太空。她說，"人試著要在太空中居住，要去月球，火星等等，這是違背上帝的旨意的。這些人為什麼不能順從上帝的旨意，乖乖地待在家裡看看電視就好呢？"

還有一個稍微早一點的版本，乃是有關反對亨利福特 T型汽車的，"假使上帝有意思要人每小時走四十英哩的話，那麼祂就會賜給他輪子，而不是雙腳。"

在某些教派的情形中，這種態度則就不是那麼好玩了。據說他們反對醫生對自然程序的"干擾"，甚至於會禁止在新生嬰孩的眼睛裡，使用硝酸銀，其理由乃是上帝"意圖"要讓罹患淋病母親的小孩，都變成是瞎子。

人的本性其實就是要去"對抗自然"。畜生活得雖然亂七八糟，但是是和自然"和諧的"；而人類卻不僅會要去改善自己，也要去改良他所處的環境。

It is a little dangerous to say simply, "God gave man a brain to use," because this kind of argument might pose a problem with respect to, say, the appendix, and also because the question is not just whether to use it but how to use it. Nevertheless, modern clergymen of most denominations are now thoroughly committed to the view that the advance of science does not imply a retreat from God.

Dr. G. Ernest Thomas, Director of Spiritual Life for the General Board of Evangelism of the Methodist Church, has written: "Religion needs science . . . The purposes of God are brought into clearer focus by every new discovery of truth which the scientist makes. . . Because religion interprets God as interested no less in the fulfillment of man's greatest possibilities as in the orderly functioning of the planets and the stars, religion honors Pasteur, Lister, Koch, Einstein, and other men of science. It recognizes the scientist as one who shares in the fulfillment of God's purposes for His world . . . I recognize that science holds the secret of a more abundant life than man has ever known." (115) (The italics are mine)

However, this does not mean that every activity of science, much less every activity of a scientist, is necessarily good, and some additional discussion of the "soul" puzzle may be useful to convince the doubtful that freezing is not sinful.

The Riddle of Soul

Besides being interesting in itself, especially in light of our later treatment of the problem of identity, a brief look at this very obscure question will serve an important purpose:

光是說 "上帝賜給人一個大腦，就是要來使用" 這是有點危險的，因為這樣的辯法，在某一方面，例如說盲腸上，就行不通了，而且因為問題不是僅在於要不要用，而是在於怎麼用。不管如何，大多數教派的現代神職人員，現在都完全地相信一個觀點，那就是科學的進步並非意涵著上帝的撤退。

　　衛理教會福音傳道部靈性生活主任，**G. Ernest Thomas**博士曾經寫道："宗教需要科學……藉著科學所產生對真理的每一個新發現，上帝的旨意才得以被帶到更清楚的焦點。……因為宗教認為上帝在對人類展現其最大可能性，並不亞於祂對行星和所有星星有次序地運轉的興趣，宗教尊敬巴斯德，李斯特，柯霍，愛因斯坦，以及其他科學家。宗教認為科學家乃是在祂的世界中，來一起分擔完成上帝的旨意的人……我深信在科學裡面，蘊藏著一個比人類目前所知道的還要更豐盛的生命的奧密。" (115) (其中的斜體字是我加入的)

　　然而，這並不表示科學界的每一個活動，更不用說一個科學家的每一個動作，就一定是好的。還有針對 "靈魂" 這個迷團的一些更深入的討論，在說服那些懷疑者來不把冰凍認為是有罪的，可能會是有用的。

靈魂的謎題

除了對其本身的興趣之外，尤其是當我們在下面章節中在對身分問題的論述有所洞悉之後，來對這個非常撲朔迷離的問題簡概地看一下，將可以達到一個重要的目的：

without denying that the soul may exist, we shall show that its definition is so vague that no one, however religious, can claim to know much about it, much less lay down moral directives about it.

In modern times, intelligent religious people apparently make little attempt to characterize the soul. It is just another Divine Mystery, rooted in faith, revelation, and especially in a kind of misty tradition. People have them; lower animals do not. (or perhaps we should say, souls clothe themselves in the bodies of Homo Sapiens, but never in those of other species.)

When are matter and spirit joined? Dr. George W. Corner says, " . . . most Roman Catholic theologians, Orthodox rabbis, and some Protestants hold that the soul is infused into the body at the moment of fertilization. To the Roman Catholic, the loss of an embryo, even if too small to be seen without a microscope, of whose existence its own mother is not yet aware, means its soul must dwell forever in limbo, outside the gates of heaven." (14)

When medical knowledge was more primitive, ideas about the soul were correspondingly different. St. Augustine and St. Thomas Aquinas are said to have written that the fetus receives its soul in the seventh or eighth week of embryonic life, which is about the time it becomes an obviously recognizable human being. (14)

在不去否定靈魂的可能存在之下，我們要突顯的就是，其定義是如此地的模糊，以至於不管一個人對宗教有多麼虔誠，還是不能說出他對靈魂到底懂多少，更遑論說要去訂定有關靈魂的道德指引了。

在當今的時代中，有知識的宗教人士，很顯然地都不會試著要去對靈魂有所界定。均將其視為僅是另外一個神聖的秘密，乃是根基於信仰、啟示、和尤其是，一種迷霧般的傳統。人類是有靈魂的，而低等動物就沒有。(或是或許我們應該說，靈魂將他們自己裹縛在現代人種 (Homo sapiens) 的身體裡，但是從來沒有在其它的物種中存在過。)

心靈和物質是什麼時候合一的呢？George W. Corner 博士說，".......大部分的羅馬天主教的神學家、東正教主教、以及某些新教徒都認為靈魂乃是在授精的時刻被溶入肉體的。對羅馬天主教人士而言，就算是胚胎小到不用顯微鏡就無法看到，其存在連他自己的母親也都無法去察覺，但是一旦失去他的話，就意味著他的靈魂會永久地漂流遊蕩在天堂的大門之外。" (14)

在醫學知識還是較原始的時候，有關靈魂的想法也相對地不同。據說聖奧古斯丁和聖湯瑪士阿奎那都曾經寫道，胎兒在其胚胎生命的第七週或是第八週時，才會接收到其靈魂，這大約是當胎兒成長到具有一個明顯的人形的時候。(14)

In 1677 Anthony van Leeuwenhoek of Delft is supposed to have regarded each sperm cell as a rudimentary embryo. His followers thought each sperm a little mannequin, itself having testes carrying tinier sperm, ad infinitum. On this basis, the German philosopher Leibniz reasoned that the first man must have carried all his descendants in his genitals, including all their myriad souls, awaiting each his turn to develop. (14)

Our main lesson from this little bit of history is that notions of soul have followed and not preceded science, and doubtless will again.

Even professional theologians have the utmost difficulty in struggling with the problem of soul. Consider the following well-meaning but pitiful effort:

" ... those who oppose the materialists insist on another kind of reality, which is not accessible to the senses . . . but only to the mind . . . a nonmaterial or spiritual world, accessible only to the reason and not to the senses . . . as when you think of numbers and geometrical figures and other abstract ideas, such as unity and freedom and love, none of which can ever be seen or touched or smelled. [To this realm] belongs man's soul . . . as well as God and whatever other spiritual beings there are." (41) (The quotations are slightly out of order.)

在 1677 年，據說荷蘭 Delft 的劉文胡克 (Anthony van Leeuwenhoek)曾經將每一個精子細胞看成為一個尚未發育的胚胎。而其徒弟則認為，每一個精子就是一個微小的迷你人，他們也有他們的睪丸，其中也有更小的精子，如此這般無限地延展下去。在這個架構上，德國的哲學家 Leibniz 推理說，第一個男人在其性器官中，一定攜帶著他所有的後代，並且也包含著他們所有多條的靈魂，等待著輪到他們成長的時刻。**(14)**

這些微小的歷史片段所給我們的主要的教訓乃是，有關對靈魂的概念，都是尾隨科學而來的，而不是在科學之前，無疑地未來也會是這樣。

就連最專業的神學家，在和有關靈魂的問題纏鬥時，也會遇到極大的困惑。請看看底下一些用心良苦，但是卻是楚楚可憐的努力：

"........那些反對物質主義者，都會堅持有另外一種現實，但其乃是不能被感官所觸及......而只能用心去感受.......一個非物質或是心靈的世界，是僅能用推理而不能用感官來進入的.........就如同當你在思考數字和幾何圖形，以及其他的抽象概念時，例如統一、自由和愛心，其中從來沒有一個是可以被看到，被觸摸到、或是被嗅聞到的。[這個範疇] 乃是屬於人的靈魂......以及上帝和所有其他任何存在的靈體的。" **(41)** (這個引述是有點錯誤的。)

That writer cannot possibly mean that God, for example, is only an abstract idea; if He were, He would be incapable of acting except through the agency of another mind. The quotation undoubtedly represents a thought, and possibly a significant one; but if so, there has been a failure of communication.

As to what the soul may be from a scientific standpoint, it is again most difficult to say. So far as I know, no one has ever devised a way to detect its existence. Since beasts, and also postulated extra-terrestrial humanoids, seem to have intelligence, personality, character, feelings, conscience, and indeed every other physical and behavioral attribute capable of detection, and yet have no souls according to religious belief, the soul seems detectable only to God.

It is also hard to see how the soul can determine identity, unless one is prepared to claim that beasts lack individuality, or that identity has a different repository in beast and man.

Perhaps, in some unclear way, the soul is not the man, but is nevertheless his most important part, somewhat as your head is not exactly you but is still the main part of you. Possibly the body can be amputated from the soul without destroying the essence, more or less as the feet can be amputated from the body without mortal damage.

這個作者可能不應該將這些都意指著上帝，例如說，僅是一個抽象的概念；如果祂果真是如此的話，那麼除非祂透過一個代理人或是另一個心靈，否則祂就無法有所作為。這個引述無疑地僅是代表著一個思維，而且可能是一個重要的思維；但是就算這樣，其中是存在有一個溝通上的謬誤。

　　至於從一個科學上的觀點來看，靈魂到底是什麼，這又是一個最難啟齒置喙的問題。據我目前所知，從來沒有一個人曾經發明出一個方法，來偵測到靈魂的存在。既然畜生，還有外星人似乎都具有智慧、個性、特徵、感覺、意識，而且的確還可以偵察到其它每一種物理上和行為上的屬性，但是依據宗教信仰來看，它們還是沒有靈魂，因此靈魂看起來似乎只有上帝才偵察得到。

　　同時要看出靈魂如何可以來決定身分也是困難的，除非我們有準備要來宣稱說動物沒有個性，或是說動物和人的身分有不同的貯存處。

　　或許以一種模模糊糊的方法來說，靈魂不代表就是這個人，但是無論如何，還是他最重要的一部份，就好像說你的頭不就是你，但是還是你的主要部分一樣。搞不好身體是可以從靈魂中切割出來的，而且還不會破壞其本質，或多或少就好像說，可以將腳切除而不會對人有致命的傷害一樣。

It is also conceivable that the soul is physically detectable after all, but only with extreme difficulty, like the neutrino. The crudeness of our observations may be at fault. There exists, of course, a substantial quasi-religious body, the Spiritualists (séances and all that), who seem to believe in a quasi-physical soul.

Some Christians, especially those literate in science, have been so impressed with the difficulties of "soul" that they advise abandoning the word altogether. Dr. Arthur F. Smethhurst, Examining Chaplain to the Bishop of Salisbury, has written: "The word 'soul' is another term, the use of which might well be abandoned in view of the ambiguities which surround it . . . If we are to reject the use of the word 'soul,' what we should substitute in place of it is probably the word 'self.' By this we must mean a self-conscious, rational human personality." (109) One suspects that the substitute word retains considerable ambiguity; but if this suggestion were widely adopted, there could be little question as to the soulfulness of the resuscitees.

Since the concept of soul in the Judaic and Christian traditions is so vague and changeable, it may not be out of order to mention the ideas of other religions and peoples. In the Shinto religion, for example, there seems to be the idea not of a soul, but simply of soul (kami). Kami refers to anything of the spirit, and it comes in variable quantities. (9)

In the Indian religions - Hinduism, Jainism, Buddhism, and Sikhism – there is belief in samsara, transmigration or reincarnation; a single soul tenants a succession of bodies. (9)

我們也可以想像得到,搞不好靈魂畢竟是可以實質地偵測到的,但是就好像要偵測到微中子一樣,只是有極大的難度。我們粗糙的觀測能力可能會造成瑕疵。當然,有一個頗大的類宗教團體,也就是靈性主義者(以及所有類似 séances 的),他們似乎相信有一種類似實體的靈魂。

有某些基督徒,尤其是那些精通科學的,已經深深領教過'靈魂"的艱澀了,因此他們建議把這個字完全地放棄掉。Arthur F. Smethhurst 博士,Salisbury 大主教的監察牧師,他寫道:"'靈魂'這個字是另外一種術語,在看到圍繞著它的各種曖昧,倒不如放棄其使用。………假使我們要排除去使用'靈魂'這個字,我們應該用來替代其位置的字,可能就是'自我'這個字了。藉此,我們一定要能夠傳達出一個自我意識的、理性的人類個性。"(109) 有人會認為這個替代的字中,也含有相當多的曖昧;但是假使這個建議已經被廣泛地採用,那麼對被復甦者還是充滿靈魂的本質,應該就沒有多少問題了。

因為在猶太教和基督教的傳統中,靈魂的觀念是如此地模糊和多變,所以去提到其他宗教和人種對此的觀念,可能也不會太唐突。例如在日本神道教中,其觀念似乎不是一個靈魂,而單純是靈魂(神,kami)。Kami 乃是指任合屬於靈的東西,而且其數量是可以變化的。(9)

在印度的各種宗教中−印度教、耆那教、佛教、和錫克教−他們都相信輪迴、轉世或是化身;而單一的靈魂是可以寄居在一系列的肉體中的。(9)

Speaking of multiple bodies brings to mind the converse idea of multiple souls. Can there be more than one to a customer? Is it possible that on clinical death the soul goes to its reward, and that if the body is revived another soul, a sort of twin-soul, occupies it? After all, we know that in the case of identical twins being born, the fertilized ovum was split into two individuals with two souls; hence either there were two souls present before the split, or else an extra one was inserted when it became necessary. A similar device might handle the death-and-resurrection difficulty, if it is deemed necessary. But we hasten to repeat that the simplest solution is to regard revival as the extension of life and not its renewal, to assume that death was not real.

The theologians in good time will decide all such questions. Or rather, several schools of theologians will each evolve a whole series of accommodations to the developing insights of science and the developing pressures of society, in the usual way.

Suicide Is a Sin

Elusive as the soul may be, Christians seem pretty much agreed that it is sinful prematurely to separate it from the body. Both murder and suicide are regarded as sinful under most circumstances, and this whether by act of commission or omission.

Physicians are generally required, by religious morality as well as civil law, to take all available measures to save life and to prolong it, even if the measures are not certain of success.

提到多重的肉體，讓我們想到多重靈魂的逆向思考。對一個收容體而言，是否有可能會有多於一個靈魂呢？是否可能在臨床死亡的時後，其靈魂退到其休息處，而假使當肉體被復活時，又有另一條靈魂跑來佔據他，而產生像是一種雙靈魂的情形呢？畢竟我們都知道，在同卵雙胞胎被生出來的時候，那個被受精的卵子會分裂成兩個個體，同時也會有兩條靈魂；因此不是在分裂之前就有兩條靈魂存在，不然就是有另一條靈魂在情況變成需要時，再穿插進去的。假使情況認定為需要的話，也可以用類似的方法，來克服死而復活的困難。但是我們急切地要再度強調，最單純的方案乃是來將復活當作是生命的延長，而不要當做更新，以便可以把死亡認為是不真實的。

神學家們在適當的時機。將可以來決定所有這些問題的答案。或是不如讓許多學派的神學家們，針對科學研發中的見解，以及社會漸生的壓力，以尋常的方法，各自來發展出一整系列相容的解說。

自殺乃是罪惡

雖然靈魂是如此地撲朔迷離，但是基督徒們似乎都相當一致地認為，時候未到時就將其從肉體分離，乃是一種罪惡。而不管是來自於蓄意的或是疏忽的，在大多數的境況下，謀殺和自殺都被認為是有罪的。

宗教的道德規範以及民法中，一般都會要求醫生採取所有可能的方法來拯救和延長生命，甚至於當這些方法，都還無法確定是一定有效的時候。

Temporary death, or clinical death with a recognized chance of resuscitation, can hardly be deemed death at all in this connection, and hence the freezers must be recognized as a probable means of saving or prolonging life.

It will then follow that failure to use the freezers is tantamount to suicide, if the decision is made for oneself, or to murder, if the decision is made for a member of your family.

Although this argument seems to me a very powerful one, not everyone will recognize it as compelling. There will be clerics on both sides of the fence.

Bishop Fulton J. Sheen, while in no way condoning mercy killings, is reported to believe that "extraordinary" medical measures should not be taken to prolong the lives of "hopelessly" ill patients. (23) Undoubtedly many other clergymen would vehemently disagree, since the line between "ordinary" and "extraordinary" measures is an arbitrary one, and the epithet "hopeless" always represents a guess. Some would say that the withholding of medical assistance. whether "ordinary" or not, does indeed constitute mercy killing.

What emerges, then, is that some few of the clergy will insist that the freezers represent an improbable means of saving life, and a disagreeable one besides, and a presumptuous and profane one as well, and will roundly condemn it. But I think the majority will take an initially cautious view, and before long will agree that failure-to-freeze represents a denial of life, and therefore of God.

在這樣的關聯之下，暫時的死亡，或是還被認定有復活希望的醫療臨床死亡，不僅絕對不能被認定為死亡，而且反而必須要去將人體的冰凍，認定為是一種拯救或是延長生命的可能方法。

依此推理，沒有去進行人體冰凍，假使這個決定是某人自己下定的，那麼這就等於是自殺；而假使這是他人為你家庭的某一成員代為下定的，那麼這就等於是謀殺。

這一個論點，雖然對我而言似乎是強而有力的，但是並不是每一個人都會將其認定為迫在眉睫的。未來在此論點藩籬的兩邊，都會有神職人員採取其立場。

雖然 Fulton J. Sheen 主教絕對不可能贊成安樂死，但是聽說他相信，我們不應該採用 "超乎尋常的" 醫療方法，來延長已經病得 "毫無希望的" 病人的生命。(23) 無疑地，一定會有許多其他的神職人士會強烈地反對，因為界定 "尋常" 和 "超乎尋常" 之間一線的方法是自由心證的，而所謂的 "毫無希望的" 一詞，也常常僅代表著一個臆測而已。不管是 "尋常" 與否，有人就可以說，停止醫療救治的確已經是在構成安樂死。

於是，就會出現某一些少數的神職人士堅持說，人體冰凍所代表著的乃是一種不可信的救人方法，一種不可能被認同的方法，之外也是一種放肆的和褻瀆的方法，因而會全然地加以譴責。但是我相信大部分的人，初期都會採取一種謹慎的觀點，而不久之後將會同意，沒有去冰凍代表著對生命的否認，因而就是對神的否定。

God's Image and Religious Adaptability

The freezer program represents for us now living a bridge to an anticipated Golden Age, when we shall be reanimated to become supermen with indefinite life spans. Indeed, even the term "superman" may eventually become inappropriate, just as a man is not aptly described as a "superamoeba" even though we evolved from a one-celled organism.

At first thought, this cannot be other than a most disturbing prospect to the Christian, Moslem, and Jew, since it seems to promise to leave Jesus, Mohammed, and Moses behind in the mists of the pre-dawn. And yet one must not underrate the adaptability of modern religions, and in fact I believe they will succeed in reinterpreting holy writ and tradition to keep pace with science and society, as they have done so often in the past.

In earlier days, there was raw conflict between science and religion. As a prominent Lutheran theologian, Dr. M. J. Heinecken, reminds us, "Whenever there was a new discovery which went counter to the traditional beliefs, the church and its leaders were quick to protest . . . Giordano Bruno was burned at the stake in 1600 because he no longer believed in a finite, enclosed universe . . . In 1632, Galileo was forced to recant his conviction that the earth revolved and not the sun. . Martin Luther did not think well of Copernicus for contradicting the cosmology of the bible . . . [and] . . . the church opposed . . . inoculation, anesthesia, birth control, and above all, the theory of evolution." (41)

上帝的形象和宗教的適應能力

對我們現在活著的人而言，當我們未來能夠被復活而成為具有無限壽命長度的超級人時，人體冰凍計劃所代表的，乃是通往預期中的一個黃金時代的橋樑。的確，到最後甚至於 "超級人" 一詞，也會變成不太恰當，就好像雖然我們都是由單細胞生物進化而來的，但是用 "超級阿米巴" 來形容一個人是不適宜地一樣。

對基督徒、回教徒和猶太教徒而言，剛開始一想到這個時，一定會產生一種極大的懊惱的景象，除此之外別而無它，因為這會讓耶穌、穆罕默德和摩西，被遺棄在破曉之前的迷霧中。然而，我們卻也不能來低估現代一些宗教的適應能力，事實上，我反而相信他們會有辦法來重新解讀其神聖的經典和傳統，而來和科學和社會並駕齊驅，就好像過去他們已經常常如此表現過了的。

在先前的世代中，科學和宗教間曾經發生過一些血淋淋的衝突。就如同一個路德會傑出的神學家 M. J. Heinecken 博士提醒我們的，"每當有一個新的發現和傳統的信仰相衝突時，教會和其領袖都會急著來抗拒……..Giorando Bruno 在 1600 年被活活燒死在木椿上，只是因為他不再相信一個有限和封閉的宇宙……..在 1632 年,伽利略被逼著去撤銷他認定地球繞日,而非日繞地球的信念…….馬丁路德也曾經沒有對哥白尼和聖經相矛盾的宇宙理論懷過好意………[而且]…….教會也曾經反對過…………疫苗接種、麻醉術、生育控制，而且最重要地就是反對進化的理論。" (41)

Happily, those days are long gone, and modern Christianity and Judaism are in the main admirably humane and forward-looking. The humanity and adaptability is wittily exemplified in two anecdotes, which came my way through Catholic friends.

The first concerns a priest who was asked by his friend, a rabbi, to contribute money to a project of the Jewish congregation, the building of a new synagogue on the site of the old. "I'm afraid," said the priest, "the bishop would not approve my helping build a new synagogue." He thought a bit, and continued. "However, there must be some expense involved in tearing down the old Synagogue, and to that I can contribute."

The second concerns a priest in a French village, in the aftermath of a battle in which invaders were successfully repulsed and one of the defenders, a Protestant soldier, died. The rules forbade burying the Protestant within the churchyard fence, and he was seemingly doomed to a lonely grave. But the good Father was equal to the occasion: he buried the soldier just outside the fence, and then labored all night until he had moved the fence, so that in the morning the new grave was in the churchyard after all. This story is not quite so funny as the first, but strikes closer to home, since it concerns adaptability with respect to customs in the disposal of bodies.

Most Christian denominations have accommodated themselves to Darwin's theory of past evolution.

很慶幸地，那些日子都已然長逝，而相當可佩服地，在當今基督教和猶太教的主流人士，都是頗具人性和前瞻性。而這種人性和適應性，很巧妙地在兩則傳說軼事中流露無遺，這些傳聞乃是經由一些天主教朋友傳到我這裡的。

第一個是說到一個神父，他被他的一個拉比朋友，要求來捐款給猶太教聚會所的建築計劃，也就是要在舊的教會的土地上蓋一個新的。神父說，"我怕主教是不會同意我去幫忙蓋一個新的猶太教會堂的。"他思考了一陣子之後，又接著說，"但是呢，要去拆掉舊會堂一定會牽扯到的一些開銷，這我倒可以捐助。"

第二個是提到住在一個法國村莊的一個神父，在一次戰役之中，入侵者被成功地擊潰了，但是有一個防衛者死了，他是一個新教徒的士兵。按照規矩，神父是被禁止在教會基園的籬笆內埋葬新教徒的，因此，這士兵看起來似乎注定要成為孤墳野鬼了。但是這個慈善的神父還是有辦法應付這種狀況：他先將這個士兵埋葬在緊靠籬笆的外邊，然後他忙了一個晚上，直到他把籬笆移了，因此在隔天早上，這個新墳墓終究還是在基園裡面了。這個故事剛聽起來好像不是那麼好笑，但是卻是相當地貼切，因為這乃是有關在處置遺體習慣方面上的適應變通能力。

大多數基督教的教派，都已經讓自己接納了達爾文針對過去進化的理論。

Dr. E. C. Messenger has written, " . . . many think there is good reason to suppose that the 'dust of the earth' of the Scriptural text need not and should not be taken to signify that the immediate source of the first human body was in fact inanimate matter. They see no reason why, on the contrary, the first human body may not have been fashioned by God from some animal organism, and this hypothesis has now been officially recognized by the supreme authority in the Catholic Church as open to discussion." (71)

Accommodation to the doctrine of future evolution, of individuals as well as the species, may be in some ways more difficult. But the same writer quotes St. Augustine as saying, "Whatever men can really demonstrate to be true of physical nature, we must show to be capable of reconciliation with our Scriptures ..." (71) This sums it up, it seems to me, rather nicely, even though it is scarcely more than a truism and leaves open the question of "should" as opposed to "could."

The problem of "God's image" in its narrower aspect should not pose too much difficulty. To be sure, man may have originally "created" God in his own image - in particular, the ancient Hebrews, I suppose, pictured God as a kind of super-goatherd but educated moderns do not seem to insist on any special physical attributes for the deity. Jesus was physically a Hebrew, but no one will assert that a Negro or an Oriental bears a more distant resemblance to God than does a Jew; or that God has any physical likeness to some of the monstrous bodies that clothe human souls. The "image" of which we speak is unquestionably a spiritual image in some sense. Maurice R. Holloway, a Jesuit writer, has said, " . . . the soul . . . is made to the image and likeness of God." (44)

E. C. Messenger 博士寫道，"……許多人認為有很好的
理由來假設聖經章節中所指 '地球塵土' 的文字，不需要也不
應該把它認定為是指說第一個人身體的直接源頭，事實上，就
是沒有生命的物質。相反地，他們看不出第一個人的身體，有
任何理由不能是上帝從某些動物中所塑造出來的，而這個假
說，現在已經被天主教教會最高權責單位，正式地認同為是可
以公開討論的。" (71)

在某些程度上，適應未來在個體上以及物種上的進化學
說，可能會較為困難。但是同一個作者引用聖奧古斯丁的話
說，"人類為了無論如何一定要能夠表現其對實體自然的忠實
度，我們一定要能夠展現出我們對聖經經文的調協能力……"
(71) 依我看來，這樣就能夠頗優雅地做出總結，雖然這幾乎不
過是一個陳腔濫調，但是卻能夠將 "應該" 的，而不是 "可能"
的問題開放擱置。

有關 "神的形象"，在較為狹義方面上的問題，應該不會
造成太多的困難。確切地說，搞不好原來是人以自己的形象
"創造出" 神的，我認為，尤其是古代的希伯來人，將神想像
成一個超級的牧羊人，但是受過教育的現代人，則對此神性，
似乎都不會去堅持有任何特殊的實質屬性。耶穌實體上就是一
個希伯來人，但是沒有人會去堅持說，一個黑人或是一個東方
人就會比一個猶太人更不像神；或是去堅持說神是具有某一實
體，像似某些怪異的軀體，來裹住人類的靈魂。我們這裡所謂
的 "形象"，無疑地在某些觀念上，乃是指一個精神上的形象。
有一個耶穌會的作家，**Maurice R. Holloway** 說過，"……靈
魂……乃是依著上帝的形象和相似所造的。" (44)

Added Time for Growth and Redemption

When we say that the human soul is made in God's image, we have only broached a topic and not capped it. Much remains to be investigated.

Clearly, the soul is capable of growth and change. Just as clearly, while it may be an image of God, it is an imperfect image. Billy Graham, Billy the Kid, and Billy-down-the-block have souls differing markedly in texture from each other and from God. Every man has the duty to seek growth and betterment, both for himself and for others.

Here, then, is another chance for the religious community to view the freezer program as a challenge and an opportunity, rather than a threat. With an extended life span, the soul has a chance to grow nearer perfection. Three score and ten simply is not enough time for respectable accomplishment, in most cases; too many jobs remain undone, too many duties undischarged, too many visions too dimly seen.

In early Christian days, the apostles expected Jesus to return in their lifetimes; later, Judgment Day was anticipated at the end of the first millennium. Now, some few sects preach an imminent Second Coming, but most Christians seem willing to agree that our earthly human history may lie mostly in the future. Likewise, in Jesus' day the average life span may have been around forty; in America now, owing to improved medical arts, including the freezer program, the average man may live for thousands of years.

成長和救贖時間的增長

當我們說，人的靈魂乃是依照神的形象所造的時，我們僅是將一個話題的洞口鑽出，而我們還沒有能將其封口。其間還留下許多應該被探討的問題。

很明顯地靈魂是會成長和改變的。同樣明顯地，雖然他可能是一個上帝的形象，但是他還是一個不完美的形象。葛培禮牧師，培禮小孩和鄰家的培禮，從他們彼此之間和從上帝看來，他們靈魂的紋理是有顯著的不同。每一個人為了自己和為了別人，都有責任來追求成長和改善。

因此，在這裡還有一個機會給信教的群體，來將冰凍計劃視為是一種挑戰和一種機會，而不是一種威脅。有了一個增長的壽命長度，靈魂就可以有較大的機會來增長為較接近完美。在大部分的情況下，七十歲的壽命簡直就是不足夠的時間，來有任何可敬可佩的成長；會有太多工作無法完成，會有太多責任無法卸下，也會有太多願景會因而黯淡不見了。

在基督教的初代，門徒們都期望耶穌在其有生之年重回；之後又預期審判的日子會在第一個兩千年來臨。現在又有一些教派，在傳耶穌即刻就會再來的道，但是大部分的基督徒，似乎都願意來接受，我們在地球上的人類歷史，大部分都是繫乎於未來。同樣地，耶穌在世的日子，人類平均的壽命可能大約是四十歲；而現在在美國，由於醫學技術的進步，其中也包含了冰凍計劃，一般人就都有可能可以活上好幾千歲。

In the case of the unconverted soul, surely the pious must welcome a chance to preserve his life and thereby extend the opportunity to save him. Letting him rot would seem to condemn his soul to Hell, whereas freezing him would allow future missionaries (or the same missionaries after their reanimation) another chance at him. I am convinced that conscientious Christians will take this argument very seriously.

Dr. Edwin T. Dahlberg, a former president of the National Council of Churches, has written something which seems relevant here: " . . . the present-day leaders of religion are beginning to appreciate the fact that science is not an enemy to be denounced but rather an ally to be welcomed as one of the *redeeming forces* in the life of mankind." (16) (The italics are mine.)

Further, we must again emphasize that the religious problems associated with increased longevity will inevitably appear whether or not the freezer program is shared by the religious. Sooner or later medical science will succeed in increasing the human life span. This has already been explicitly recognized by Christian writers.

Dr. Gene Lund, professor of religion at Concordia College, is one. "Who knows but what a decade or two hence the average man will comfortably reach an age of one hundred years - at least." (63) He goes on to say, "But science does not have, and never will have, the power to eliminate death."

對還沒有皈依的靈魂而言，那些虔信者一定會歡喜有一個機會，來保存他的生命，因此可以來延伸拯救他的時機。如果任其腐壞，似乎就是判定他的靈魂下地獄一樣，然而如果將其冰凍起來，未來就可以有對他傳道的機會 (或是在他們被復活之後，對他們傳同樣的道)。我是相信有良心的基督徒，將會非常嚴肅地面對這個論點的。

全國教會協會的一個卸任下來的主席，**Edwin T. Dahlberg** 博士曾經寫過一些似乎與此相關的東西：".........當今宗教界的領袖都開始體認到一個事實，那就是科學不是一個該被譴責的敵人，反而是一個該受歡迎的盟友，來在人類的生命中充當一個*救贖的力量*之一。" **(16)** (斜體字是我附加的)

我們進一步必須要再度強調的，不管冰凍計劃是否有被宗教界認同，有關延長壽命的宗教問題，不可避免地將一定會浮現。遲早，醫療科學將會成功於增長人類生命的長度。這個已經被一些基督教的作家公開地認定了。

Concordia 學院宗教系教授 **Gene Lund** 博士就是其中的一個。"誰會知道距今僅僅一二十年之後，一般的人將輕輕鬆鬆地就至少可以活到一百歲的年齡。" **(63)** 他又繼續說，"但是科學沒有，而且將永遠沒有能力來消滅死亡。"

In other words, the Christian can expect, and welcome, the prospect of increased longevity, and cannot set any limits on it. At the same time, permanent death will surely come some day, however long deferred; science can give us indefinite life, but not literal immortality, not mathematical eternity. Hence the freezer program, if we take a sufficiently long view, is not so radical after all, but merely another incident in the cosmic drama. The freezer program is merely a medical means which will allow the present generation to share the longevity which our descendants will have in any case.

Conflict with Revelation

Some Protestant denominations, in particular, make much of Revelations in the New Testament, and can be expected to oppose a program that does not seem to fit their view of God's plan for history. But Christianity as a whole is unlikely to make a stand on this issue, because the pertinent passages are so obscure and there is so much disagreement about their meaning.

For example, Dr. Merrill C. Tenney, writing about the Millennial Kingdom, tells us: "There are three main interpretations of this passage. (20: 1-6) The post-millennial view looks upon the Millennium as a period closing the conquest of the world by the preaching of the Gospel . . . His Kingdom comes. At the end of an indefinite period of peace and righteousness, He will return to judge the living and the dead, and the ages of eternity will begin.

換句話說，基督徒可以來預期和歡迎壽命增長的前景，但是卻不能對它訂定任何的限度。同時，永久性的死亡有一天將一定會來臨的，只是會延後許久而已；科學可以給我們不定期長的壽命，但是不是字面上的不朽，也不是數學上的永恆。既然，假使我們採取一個較長遠的觀點來看，冰凍計劃畢竟不是那麼地極端，而僅是宇宙戲劇中的另一個事件。冰凍計劃只是一種醫療的方法，讓現今的這一代人可以享受到我們的子孫將一定會擁有的長壽。

和啟示錄的衝突

　　特別是改革宗教會中的某些教派，對新約聖經中的啟示錄非常重視，因而以他們的觀點來看，對一個似乎不符合上帝規劃中的歷史計劃，可以預期到，他們一定是會反對的。但是就基督教整體而言，他們是不太可能會堅持這個問題的，因為其間與此有關的經文，都是如此地曖昧，以致於對其真義的爭吵會多如牛毛。

　　例如 Merrill C. Tenney 博士，他在寫有關千禧國度中告訴我們："相關於這段經文 (啟示錄 20 章 1-6 節) ，有三種主流的解經。後千禧年者 (post-millennial) 的觀點，將千禧年看成是藉由福音的傳教，而導致世界爭戰休止的一段時間．......祂的國度降臨了。在到達一段沒有限期的和平和公義的末尾時，祂將回來，審判活人和死人，接著永恆的世代就將開始。"

"The amillennial view treats the thousand years as wholly figurative . . . There will be no outward and visible reign of Christ on earth until after the judgment.

"The premillennarian view holds that Christ will return to earth to abolish all outward opposition, that He will establish here an outward visible Kingdom lasting one thousand years more or less ..."(115)

There is certainly ample room here for the view that the freezer program is part of God's plan.

It is interesting to remark the accommodation that has been made by certain modern Jews in Israel with respect to the prophecies of Messiah. Christians, of course, believe Jesus was the Jewish Messiah, although He did not seem to fill the bill well enough to convince many Jews; some modern Jews still expect Messiah to appear; but a substantial body of modern Jewish opinion, if I understand correctly, holds that the State of Israel embodies the concept of Messiah, with no haloed individual to be expected.

In a vaguely similar way, then, perhaps it is even conceivable that the freezer era -- if it develops into an age of brotherly love and a living Golden Rule, as I believe it will -- may be accepted by some as the embodiment of the Millennium.

"不相信千禧年者的觀點，認為一千年完全是比喻性的……將不會有基督在外觀上和實質的統治，直到在大審判之後。"

"前千禧年者的觀點，相信基督將會回到地球，來清除所有外在的反對勢力，而祂將會在此建立一個外觀上可見的國度，並且會持續一千年左右……" (115)

在這裡面，當然地還留有充分的空間，來容納冰凍計劃乃是上帝規劃中的一部份的觀點。

針對彌賽亞的預言，很有趣地，我們注意到某些現代在以色列的猶太人，已經做了妥協。當然，基督徒是相信耶穌就是猶太人所謂的彌賽亞，雖然祂似乎沒有具備足夠的條件，來說服許多猶太人；甚至有某些現代的猶太人，還在期待彌賽亞的出現；但是假使我的理解正確的話，現代猶太人意見中的一個相當大的主體認為，以色列這個國家，就是彌賽亞這個概念的具體呈現，而是不需要去期待一個頭戴光環的個人。

因此，以一個曖昧的類比方法來講，假使冰凍紀元，可以發展成為一個四海皆兄弟的愛，和一個生存的黃金定律的世代，那麼就如同我相信的，它將可能被某些人接受為就是千禧年的具體展現。

The Threat Of Materialism

The pious have long been afraid of the know-it-all attitude sometimes engendered by science; they decry the loss of the sense of wonder at the mysterious universe. In this connection, Dr. Gene Lund has quoted a verse attributed to Peter Marshall:

> *Twinkle, twinkle, little star -*
> *I know exactly what you are:*
> *An incandescent ball of gas,*
> *Condensing to a solid mass.*
>
> *Twinkle, twinkle, giant star -*
> *I need not wonder what you are,*
> *For seen through spectroscopic ken*
> *You're helium and hydrogen.*

But whatever the effect of scientific advancement on the man in the street, the scientists themselves usually have a very lively sense of wonder, if not of awe. Many of them, including some of the greatest, have also been deeply religious - e.g., Copernicus, Galileo, Kepler, Boyle, Newton, Priestley, Faraday, Eddington, and Pasteur, as well as a host of moderns.

Does the freezer program, then, really threaten the existence of the mass of the people in that it will become hopelessly secular and materialistic?

物質主義的威脅

　　某些虔誠的人士，長久以來就一直害怕到某個時候，由科學所導致的一個無所不知的情境；他們譴責人們對神秘宇宙發出驚歎的感覺，已然喪失。在這個關節點上，**Gene Lund** 博士有引用了來自 **Peter Marshall** 的一段詩句：

> 一閃一閃小星星，
> 我知到底你是啥：
> 一顆氣體熾熱球，
> 冷凝成一固態體。

> 一閃一閃大星星，
> 我不好奇你是啥，
> 因透光譜視野看，
> 你是氦氣和氫氣。

　　不管科學進步對市井人士有什麼影響，但是科學家們自己，通常就算不致於驚歎的話，至少也都會有一個非常活躍的好奇感。他們之中有許多個人，其中包括某些最偉大的，也都是信仰非常虔誠的－例如哥白尼、伽利略、刻卜勒、波義耳、牛頓、普利斯特例、法拉第、愛丁頓、以及巴斯德，還有一大堆現代的科學家。

　　那麼，冰凍計劃是否真的會威脅到人類群體的生存，以致於變成不可救藥地世俗和物質呢？

The answers are fairly obvious, but let us display them anyway, after devoting a few words to the always bothersome question of definitions.

A "materialist," as the word is often used in a derogatory sense, is someone who is blind to things of the "spirit"; in extreme cases it means someone who is obsessed with wealth and sensuality and does not appreciate the values in art and in human relations. As I prefer to use it, however, it merely means someone who is not a dualist, someone who conceives of the universe as unitary, without any dichotomy between "matter" and "spirit."

"Religion" is much harder to define. According to the Rev. M. R. Holloway, "Religion. . . consists in that act by which man worships God, subjecting himself to Him." (44) But this definition seems much too narrow.

One of the organized religions, Buddhism (at least in some of its forms), does not even concern itself with a deity! Millions of Buddhists have religion but no God. Furthermore, many writers have acknowledged that Soviet communism has essentially the character of a religion. Seeking the common elements, we can probably say that the essence of religion lies primarily in extreme dedication, and secondarily in fellowship.

It is plain enough that man can get along without religion in the narrow sense - or at any rate some men can. Many men get along without it in America today, just as many got along without it in classic Athens, including great and good individuals.

其答案是頗明顯的，但是在先費一些口舌來解說一些常常會令人困擾的定義問題之後，我們還是會將之呈現的。

所謂一個 "物質主義者"，因為這個字眼常常是用在貶損的用意上，所以往往是指某人對屬於 "精神" 的事物是盲目的，而在某些極端的情形下，乃是指某人沉迷於財富和感官，而沒有去體驗存在於藝術和人際關係中的價值。然而，就我較偏愛的用法而言，它僅是指某人不是一個二元論者，把宇宙想成是單元的，而沒有任何 "物質" 和 "精神" 的二分想法。

"宗教" 就比較難以定義了。依據 M. R. Holloway 牧師，"宗教.......其中包含著人類崇拜神的行為，將自己臣服於祂。"**(44)** 但是這個定義看起來似乎太過於狹隘。

組織性的宗教之一，佛教 (至少在其中的某些型式中)，甚至於連一個神都沒有牽扯到,好幾百萬的佛教徒是有宗教信仰，但是卻沒有神。況且有許多作家都認知到，蘇俄的共產主義，本質上就具有一個宗教的特徵。為了要尋求其共通的要素，我們可能可以說宗教的要質，主要地乃是在於其極度的虔信，而次要地則是在於其團契的行為。

夠明顯地，人是可以在沒有定義狹隘的宗教下活得好好的，或是至少有一些人是可以的。在當今的美國，有許多人是活著沒有它的，就如同在古代的雅典，有許多人都沒有它一樣，其中單然包括了許多偉人和好人。

But whether many people could get along indefinitely without some kind of dedication and fellowship is another question, and the answer is probably negative.

It follows that the church as an institution is in no danger. It offers a formal dedication which fills a deep-felt want. It offers - even without Bingo – a warmth of fellowship hard to find elsewhere. Like everything else pertaining to man, the churches will change, but they will not die.

Perspective

The religions are willingly and foresightedly undergoing a continuing process of reexamination and adaptation in light of new discoveries and new capabilities, of which the freezer is only one. Precedent already exists for regarding preservation and reanimation of seemingly dead people as routine medical procedure, aimed simply at prolonging life. The religious problems, if any (as well as the economic and social problems) related to extended life have long existed, and will continue to grow, with or without a freezer program. When the freezer program gains momentum, religious people, except in scattered instances, are not likely to be left behind.

但是，至於許多人是否可以一直活著，而沒有某些類型的虔信和團契，這則是另外一個問題，而其答案可能是否定的。

因此，可以說宗教這一種組織是不會被危害到的。它提供了一種正式的敬拜，可以滿足一種內心深處的渴望。就算沒有賓果遊戲，它也提供了一種在其它地方難以覓得的團契的溫暖。就像其它所有和人相關的事物一樣，宗教是將會有所改變，但是他們將不會消逝的。

透析

在新的科學發現和新的可能的光照之下，而冰凍計劃只是其中之一，宗教界都很願意，而且都蠻有前瞻力地，來在進行一個連續性的檢視和適應的過程。有關把看起來似乎已經死亡的人的保存和復活，當作一種醫療程序上的例行公事中，在此之前，已經有了專門針對壽命延長的先例了。這些宗教上的問題，假使是和壽命延長有關係的（也包括經濟上和社會上的問題），不管有沒有冰凍計劃，其實是早就存在了的，而且將會繼續增加。當冰凍計劃蓄積了足夠的動力時，除了一些零星少數的特例外，宗教界人士也不可能會被遺忘在後面的。

CHAPTER VI

Freezers and the Law

冰凍人和法律

CHAPTER VI

Freezers and the Law

Even though our Supreme Court is sometimes accused of radicalism, jurisprudence in general is very conservative. In fact, some jurists face rigidly backward; they don't care where they're going, but only want to know where they've been. They are perpetually astonished that tomorrow always arrives. But they, like their more progressive colleagues, may as well face up to the fact that there really is a future, and that it is more comfortable as well as more dignified to walk into it than to be dragged in.

Not only the bodies of the frozen must be protected, but also their property; and not only their property, but also their rights. Remember Ralph Waldo Emerson: "For what avail the plough or sail or land or life, if freedom fail?" The defender of the status of the frozen, as of us all, must be the law.

The law indeed, but what law? Why, the law that will be shaped in the usual way, in the legislatures and more especially in the courts, by test, re-test, compromise, and evolution. While its outlines are still dim, we can look at some of the obvious problems, and conjecture about solutions.

第六章

冰凍人和法律

雖然有些時候，我們的最高法院往往被指責為激進主義，但是畢竟整個法理體系，一般而言，還是非常地保守。事實上，有某些法官還是頑固地引頸過往；他們不管未來是要走向何方的，而只想知道他們曾經到過哪裡。他們一直都是活在明天竟然是會來臨的驚訝中。但是，他們倒不如和他們其中較前衛的同事一樣，面對確確會有一個未來的現實，這樣他們也能夠較舒坦地，並且較有尊嚴地，來邁入未來，而不是被拖曳進去的。

不僅被冰凍者的身體必須要被保護，他們的財產也要被保護；而且不僅是他們的財產要被保護，他們的權利也要被保護。記得 Ralph Waldo Emerson 曾經說過，"假使失去了自由，那麼有耕犁或是船帆或是土地或是生命又復何用呢？" 被冰凍者地位的保衛者，就好像我們所有人的地位一樣，一定是要靠法律的。

的確是要靠法律，但是到底是什麼法呢？為此，這個法律將會依著慣常的方法，在立法機關，以及特別是在各個法庭中，經由考驗，再考驗，妥協，演變，而來被定讞。雖然，目前其輪廓仍然是模糊的，但是我們還是可以先來檢視一些明顯的問題，然後推測其解決方案。

Freezers and Public Decency

To begin with, there will be an attempt to fit the freezers into the structure of laws governing the disposal of bodies and the operation of cemeteries, mausoleums, and home crypts. Conceivably, this attempt might cause some localities to try to outlaw freezers altogether; but the advantage would seem to be with those backing the freezers.

Present laws in general seem to give priority to the wishes of the deceased and of the next of kin, subject to the community interest with respect to health hazards, property values, and common decency. Courts of equity have power to settle controversies as to the burial of the dead, the care of their remains after burial, and the preservation of the place of interment from wanton violation or unnecessary disturbance. (73)

There is legal precedent to allow unusual treatment of bodies. In Seaton v. Commonwealth, 149 Ky. 498, the defendant buried his child in a wood in a paper box, without religious ceremony, yet the court held that no criminal action would lie. (73) Michigan law states that the next of kin ". . . can bury the corpse in any manner he sees fit, so long as it does not outrage public decency or amount to a public nuisance." (73) But a disposition permit is required.

Further, the burden of proof will apparently rest on those opposing the freezers: "An unlawful, improper or dangerous establishment . . . may be enjoined, but not at the suit of one . . . who cannot show the likelihood of positive and substantial damage." (72)

冰凍人和公眾禮儀

一剛開始，一定將有人會嘗試將冰凍計劃，納入目前在管制遺體處理和墓園，靈骨塔和家族骨窖營運的法律結構中。可想而知地，這個嘗試可能會導致某地區的人，會想要來讓冰凍計劃變成完全地非法；但是，勝勢看起來似乎還是落在支持冰凍計劃的這一方。

一般而言，當前的法律在考量有關健康的危害，財產的價值，和公眾的禮儀等的社群利益之下，似乎對亡者和其近親的意願有給以較高的優先權。有關亡者埋葬，葬後遺骸照料，以及保護其收容處所免於被肆意的入侵或是莫需有的騷擾，公正的法庭有權力來處置這一些爭端。(73)

針對遺體不尋常處置的容許，已經有法律上的先前判例。在 Seaton 對抗 Commonwealth 的判例中，149 Ky. 498，被告者將其小孩放入一個紙盒中，然後在沒有舉行任何宗教儀式下，將其埋葬在一個樹林裡，但是法庭認為沒有任何根據，來判定其有任何犯罪的行為。(73) 密西根州的法律表示直系親屬 "⋯⋯可以以任何他認為合宜的方法，來埋葬遺體，只要這不會觸犯到公眾的禮儀，或是造成一個公眾的厭惡。" (73) 然而，還是要去取得一個處置的許可證。

還有很明顯地，舉證的責任還是會落在反對冰凍計劃的人的身上："一個違法，不恰當或是危險的確立⋯⋯可能是值得慶幸的，但是，卻不是當在訴訟中一個人⋯⋯不能提出實質的和可觀的傷害的可能性下。" (72)

If some locality decides a freezer is an outrage or a nuisance and orders burial, the relatives of the frozen will no doubt be able to obtain a temporary restraining order against enforcement, since time would be vital only to the frozen. If lower court decisions should be adverse (which is not easy to imagine), then probably the issue could be taken to the United States Supreme Court on the question of "equal protection," granted in the Fourteenth Amendment.

If, for a time, freezing in some localities is legally too difficult, then many people will leave those localities.

Definitions of Death; Rights and Obligations of the Frozen

The only definition of death acceptable to a biologist is that of Dr. A. S. Parkes: "Death is the state from which resuscitation of the body as a whole is impossible by currently known means." (110) Implicit in Dr. Parkes' definition is nearly the main thesis of this book: that if we use extreme freezing to prevent deterioration, sooner or later "currently known means" will be adequate, and the body will no longer be regarded as dead. The present legal definition of death, effectively, is simply any condition discouraging enough to induce the attending physician to sign a death certificate. Usually this means "clinical death" - cessation of breathing and heartbeat - but not necessarily, since artificial respiration, heart massage, or other measures may be indicated.

假使，某一地區認定一個冰凍計劃乃是一種侵犯和噁心，因而命令其要埋葬，此時要被冰凍者的家屬，無疑地，將可以先來取得一個暫時反對強制命令的緩行令，因為只有時間對被冰凍者而言是重要的。假使，低級法庭的判決是不利的 (這是難以想像的)，那麼這個爭議可能就可以依據美國憲法第十四條修正條例中，有關 "保護平等" 的質疑，而來上訴到美國的最高法院。

　　萬一，暫時在某些地區，冰凍計劃要合法化太困難的話，那麼有許多人將會搬離開那些地區。

死亡的定義：被凍者的權利和義務

　　對一個生物學家而言，唯一可以接受的死亡定義，就是如 A. S. Parkes博士所說的："死亡，乃是用目前所知道的方法，不可能來完整地救活一個人的狀態。" (110) 隱藏在Parkes博士的定義中的，幾乎就是本書的主要論點：那就是，假使我們用極低的冰凍，來避免腐敗，那麼遲早這個 "目前所知道的方法" 將會變成夠用，因此這個人就不再被認定為死亡了。目前法律對死亡的定義，實質上就是，只要有任何狀況，無望到會讓照料的醫師願意來簽出一章死亡證明書。通常這就是所謂的 "臨床醫療死亡"－呼吸和心跳的停止－但是其實是未必的，因為有人工呼吸，心臟按摩，或是其他措施的成功可能可以來證明。

When we quick-freeze a freshly dead corpse, we have someone who is thoroughly dead by current criteria, but who has potential life in almost the same way as a drowning victim who may be helped by a respirator. This is something new in the world of jurisprudence, and account must be taken of it.

When suspended animation becomes feasible, some will choose to be frozen alive, making their journey to the future first-class, perhaps with stop-overs along the way to check on conditions. While in the freezer, such a person will not be dead by Dr. Parkes' definition. Yet his active life will be only potential; he will be thoroughly inert and will require a special kind of legal status and protection, just like an actual cadaver.

Heretofore a corpse has had in itself neither rights nor obligations; now it will have both. His rights will include protection of his body and of his property, governmental supervision of the freezer and of his trust funds. His obligations will include the duty to pay taxes out of his funds and property and to submit his estate to regulation. Furthermore, the manner of his previous life and of his death may affect the nature of his privileges and duties after resuscitation.

Perhaps the law will come to recognize three classes of people in addition to the active nuisances: those in suspended animation, those frozen after death, and those who are thoroughly dead because they were burned up, well rotted, lost at sea, or otherwise considered poor bets. We can anticipate some sticky lawsuits questioning the categories assigned in particular cases.

當我們在快速冰凍一具剛剛死的屍體時，其時我們乃是有了一個，以目前的標準來講，是完全死亡的人，只是他幾乎像是一個被水溺斃的受害者一樣，還有潛在的生機，可能可以用呼吸器來幫助他。這對法理體系的世界而言，乃是新鮮的事情，而且是一定要將其列入考量的。

　　當活體暫眠的技術變成可行時，有些人可能將會選擇來在活著的時候，就被冰凍起來，使得他在前往未來的旅途，搭的是頭等艙，或許沿途會有幾次的暫停，來觀察情勢。當其在冰凍櫃時，依據 Parkes 博士的定義，這種人將不是死的。然而他的生命活性，將僅僅是潛在的能力；他將完全是鈍惰的，如同一具實質的屍體一樣，將會需要有一種特殊種類的法律身分以及保護。

　　直到此時之前，一具屍體，在其本身根本就沒有任何的權利與義務；而現在他將擁有兩者。他的權利將包括對他的身體，和對他的財產的保護，以及政府對他的冰凍器和他的信託基金的監護。他的義務將包括用他的基金和財產來納稅，以及將他的產業納入管制。還有，他的前一個生命和他的死亡的狀態，可能會影響到他在被復甦後的權利和義務的實質。

　　或許法律除了活著的麻煩製造者之外，將會認定另外三種不同種類的人：那些活體暫眠的人，那些死後被冰凍的人，以及那些徹徹底底死亡的人，例如被燒焦了的，完全腐化了的，葬身遺失海底的，或是其它所之的不幸。我們將可以預期到，會有一些棘手的訴訟，來質疑某些特殊案例所被判定的類別。

Life Insurance and Suicide

Will a frozen individual be dead enough for the beneficiaries to collect his life insurance? There will usually be two beneficiaries, the next of kin and the corporation handling his freezer and trust funds.) At first thought, the answer seems obvious: since he died in the ordinary course of events, the actuarial basis of his insurance is unchanged, hence the insurance company has suffered no unusual loss and should pay off. But on second thought, things are not quite so simple.

Will not the suicide rate increase? It seems likely that some people not desperate enough to face permanent death might reach the point of choosing premature death followed by freezing, hoping to awaken to find vanished problems and a new life.

This particular problem seems easy to solve. At present, the insurance companies typically pay off on suicide if it is not within two years of the date of issuance of the policy. In the freezer era, the insurance companies will either insert a straight-ban suicide clause or use some kind of sliding scale based on experience. A few enterprising characters may try to camouflage suicide as accident, but this will not be an easy trick, remembering that the body must not be badly damaged and must be quickly available for freezing. Falling out of windows or under subway trains won't do.

壽險和自殺

　　一個被冰凍的個人，是否死得夠徹底，因而其受益人可以申請他的人壽保險的理賠呢？(這通常將會有兩個受益人，他的第一近直系親屬，以及處理他的冰凍和信託基金的公司。) 剛一想到此時，這個答案似乎非常明顯：因為他是經過事件的一般過程而死亡的，所以他的保險的精算基礎是不變的，既然其保險公司並沒有蒙受任何不尋常的損失，那就應該理賠。但是再度思考時，事情並不是那麼地單純。

　　自殺率是否將會因此而增高呢？某些人看起來似乎會在當其還沒有危急到要面臨永久的死亡之前，就可能已經到達一個田地，會去選擇冰凍之後的提早死亡，來期待在其被復甦之後，發現所有的問題都消失了，而且有一個嶄新的生活。

　　這個特別的問題，似乎是容易解決的。在目前，假使這不是發生在簽發保險契約日期的兩年之內的話，保險公司通常都是會對自殺給以理賠的。這些保險公司，在冰凍人的世代中，將不是會插入一個直接禁止自殺的條款，就是會使用一些基於經驗值的某些變動衡量尺度。有一些具有企圖心個性的人，可能就會試著把自殺喬裝成意外，但是這種詭計將不是那麼地容易，請記得，身體一定要沒有太嚴重的損傷，而且必須要很快地就可供進行冰凍才行。要是從窗戶跌下，或是被地鐵輾過，都是行不通的。

Suicide has always been illegal. The Earl of Birkenhead tells us that in eighteenth century England, on at least one occasion, attempted suicide was punished by hanging the wretch! (6) Served him right, no doubt. Now actual suicide will become punishable as well, perhaps by imposing fines on the estate of the frozen, who will then awaken poorer than he had hoped. We can't have people just sneaking off, shirking their responsibilities.

But the illegality of suicide will have to be carefully reviewed, for clearly there can be extenuating circumstances. If some poor devil is wasting away with an excruciating cancer, he may decide to kill himself and be frozen -both to spare himself the terminal agony and to freeze his body in better condition, as well as to save further hospital bills. Similar remarks could be made about various kinds of unfortunates with crippling deformities. The legislatures will no doubt set up standards, and the courts will issue suicide permits.

We may also note the need for a new word to distinguish destructive suicide from self-inflicted temporary death. Maybe we could call it "sui-term" or "sui-kaput," to indicate that one has not merely killed himself, but ended himself.

Mercy Killings

Closely related to the problem of suicide is that of mercy killing.

自殺，一直以來都是違法的。**Birkenhead** 伯爵告訴我們，在十八世紀的英格蘭，在至少有一個案件中，企圖自殺是會判這罪犯吊死之刑的！**(6)** 罪該應得，毫無疑問。現在實質的自殺也將會是可罰的，或許可以對被冰凍人的家產加諸罰款，他在被復甦之後，將會比他期望中的還要窮。我們是不能讓人可以偷雞摸狗，逃避他們的責任的。

但是有關自殺的不合法性，將必須要被重新小心地檢視過，因為很清楚地，其中可能有可被原諒的情境。假使某一個貧窮的傢伙，浪費生命在一種極為痛苦的癌症中，他可能就會決定來自殺，然後被冰凍，這不僅可以讓他自己免除絕症的折磨，以及能夠在較佳的狀況下，冰凍他的身體，同時也可以省下進一步的醫院費用。同樣的說辭，也可以用在罹患有局部癱瘓變形的各種不同種類的不幸的人身上。立法單位將一定會豎立出一些標準，而且法庭也將會發出自殺許可令。

我們可能也要注意到，必需要有一個新字眼，能夠分辨出毀滅性自殺和自我施加的暫時死亡。或許，我們可以稱之為"自滅，**sui-term**"或是"自了，**sui-kaput**"，來表明這個人不只是殺了他自己，並且還是完全地終結了自己。

仁道致死

和自殺問題息息相關的，就是仁道致死的問題。

Under what circumstances, if any, will the next of kin be allowed to decide whether a blighted life should drag on or be mercifully frozen? Under what circumstances will the courts make this decision?

If an aged parent is in an institution with his mental faculties largely gone, is it right to keep him there? Would he not be better off frozen before his brain deteriorates further? And cannot the family's financial resources be better used to provide him a trust fund than to support him in a sanatorium?

What about a hideously deformed and defective child, as in a severe case of cretinism? Must his life drag on, in the usual way, at bitter emotional expense? Would not early freezing be a true mercy? Some will say that if we freeze all the cretins, there will be no way to study cretinism. Others will go further and say that if we freeze all corpses, the medical students will have no one to slice up in the freshman course in Gross Anatomy. However, there will probably be enough such objectors to save the situation, for surely they will volunteer their carcasses to the medical schools!

The painful problem of deformed and defective children is not one of negligible proportions. According to Jane Gould, "In all, roughly three newborn infants out of a hundred are seriously abnormal." (35) Most of these, of course, will not be considered for early freezing; they will either die early natural deaths, or will be cured, or can be helped to lead lives not too pitifully far from the norm.

假使有的話，在什麼情境下，直系親屬可以被容許來決定一條垂死的生命，是否應該要讓其苟延殘喘，還是要被仁慈地冰凍呢？在什麼情境下，法庭將會做這個決定呢？

　　假使一個老邁的雙親是在一個安養院裡，而其神智官能大部分已經喪失，繼續把他安置在那裡是對的嗎？在其大腦還沒有進一步退化之前，將他冰凍起來，不是較好嗎？而且把家族的財務資源用在提供他一個信託基金，不是會比耗在一個精神療養院來得好呢？

　　對一個已經醜陋地變形和有缺陷的小孩，例如一個嚴重癡呆症的案例，又該怎麼辦呢？難道他的生命必須一定要付上這麼痛苦的感情代價，來蕭規曹隨地拖延下去嗎？提早冰凍不就是真正的仁慈嗎？有人將可能會說，如果我們將所有的白痴兒冰凍，那麼就沒有辦法來研究白痴症了。其它的人可能會更進一步說，假使我們將所有的屍體冰凍的話，在其新鮮人大體解剖的課程中，醫學院的學生就沒有一個可以用來切割的了。無論如何，總是會有夠多反對這事的人，可以來免除這種困境，因為他們將一定會把他們的屍體捐出來給醫學院的！

　　變形和有缺陷小孩的這個痛苦的問題，絕對不是一個可以輕易忽視的。根據Jane Gould，"總括來看，大約每一百個新生的小嬰孩中，就有三個是嚴重地畸形。" (35) 當然，其中的大部分將不會被考慮來提早冰凍；他們不是將會提早死於自然死因，就是會被治癒，不然就是可以被幫助來過一個不會離正常人太可憐地遠的生活。

But consider, for example, the worst cases of cerebral palsy. According to Jessie S. West, in the United States in 1954 there were around a half million victims of this disease. Many had normal intelligence, although the affliction produced symptoms such as facial grimacing, drooling, and unintelligible speech which might make them seem subnormal to an uninformed observer. But many had serious mental deficiencies, and in fact 13 per cent were considered uneducable. (127)

At present, we properly do not countenance euthanasia for this 23 per cent, even though they may be suffering and even though there is a heavy emotional and financial burden on the other members of the family. But will not the situation be different when freezers are available?

Some will insist that we cannot end life for any reason, let alone for reasons of cost and convenience. But in fact we have always sold lives, and sometimes rather cheaply, in peace as well as in war. Consider, for example, the annual American traffic death toll - around thirty thousand, I think. We could certainly save several thousand of these, merely by doubling the police traffic detail in every city, or by making all vehicles carry speed governors, etc. But we do not want the expense or inconvenience of saving these lives; we make a cold-blooded calculation, and let them die.

Certainly there is an extremely important difference between traffic deaths and mercy killings. In the former case the victims are not known in advance, and we all take our chances. Nevertheless, life does have its price, and the freezers introduce a profoundly important new element.

但是想一想，例如像是一些最嚴重的腦麻痺病例。根據Jessie S. West，1954 年在美國，就大約有五十萬這種疾病的受害者。雖然罹患此病會產生一些症狀，像是顏面扭曲，流口水，以及胡言亂語，從一個不知情的旁觀者而言，這可能使得他們看起來似乎是低能的，但是其中很多人是具有正常的心智的。然而畢竟還是有很多是有很嚴重心智不足的現象，而且事實上，其中有百分之 23 是被認為是不可受教的。(127)

　　在目前，儘管他們可能是在受苦受難中，儘管這對家庭中的其它成員，是一個沉重的精神上和金錢上的負擔，我們恰當地沒有支持這個百分之 23 的人去進行安樂死。但是當有了冰凍計劃後，這個情勢不就將會有所轉變嗎？

　　有些人將會堅持，我們不能夠以任何一個理由來終結生命，更不用說是以花費和方便為藉口的了。但是事實上，不管是在平時或是戰時，我們一直都在出賣生命，而且有時候是在賤賣。想看看，例如美國每年的交通事故死亡人數，我想大約就有三萬人以上。在每一個城市中，只要靠著加倍警察的交通勤務，或是藉著強制所有的車輛都配有速度控制器等等，我們就一定可以拯救其中好幾千條的人命。但是我們不想要有這些花費和不方便，來拯救這些生命；我們做了一個冷血的計算，而任由他們去枉死。

　　當然在交通事故死亡和仁道致死之間是存在有非常重要的區別。前面的案例中，受害者是誰是不得預先知道的，而且我們每一個人都有機會去遭遇到。畢竟生命是寶貴的，而且冰凍計劃則引進了一種極度地重要的新要素。

One cannot evade his responsibility by speaking of "God's will." The failure to act also constitutes a decision. When the judge is pondering the case and searching his soul for right, let him ask himself this question: if the child were already frozen, and it were within my power to return him to deformed life, would I do so? If the answer is negative, then probably the freezer is where he belongs.

Murder

In the new era, the heinousness of the manslayer's crime may depend not only on the motives and circumstances, but also on the degree of damage to the body.

My grandfather used to say there are two kinds of lazy -- "lazy" and "stinking lazy." Society may now distinguish between plain murder and sloppy murder. If the victim is doused with gasoline and ignited, or ground up in the garbage disposal, or hidden in a swamp and left for the alligators, this is sloppy murder. But if he is merely shot through the heart and quickly found and frozen, then this is a more civilized kind of murder.

The punishment of murder will have to be reviewed. Should it fit the crime? Should one who destroys his victim be himself scattered to the winds? Should one whose victim can be frozen be himself frozen? In those states which do not use capital punishment, can freezing be substituted for life imprisonment in some cases?

人不能夠藉著說，這是 "神的旨意" 來逃避他的責任。沒有去做某事，也構成是另一種抉擇。當法官在思考這個案子，並且搜尋他的靈魂要追求正義時，請他問他自己這個問題：假使這個小孩已經被冰凍，而且在我的能力之中，有辦法將他還原到其本來畸形的生命，我是否會如此做呢？假使答案是否定的話，那麼冰凍櫃可能就是他應該歸屬的地方。

謀殺

在新的時代中，殺人犯罪行的惡質程度，可能不僅是會和其犯罪動機和情境有關，同時也會和受害者身體損壞的程度有關。

我的祖父過去常說懶惰有兩種一"懶惰" 和 "又臭又懶"。社會現在可能已經會辨別純粹的謀殺和粗劣的謀殺了。假使受害者是被澆了汽油，然後點燃，或是用垃圾處理機絞碎，或是被丟入一個沼澤，然後任由鱷魚吞吃，這就是粗劣的謀殺。假使，他僅是被一槍穿過心臟，然後很快就被發現，而被冰凍起來，那麼這就是一種較為文明的謀殺。

針對謀殺的刑罰，將必須要被重新評估過。是否應該要和其罪過相當呢？那個將其受害者摧毀的人，是否也應該被碎屍萬段，灑向空中呢？受害者如果還可以被冰凍的話，那他是否也該被冰凍呢？在那些不准實施死刑的州，在某些情形下，是否可以使用冰凍來替代無期徒刑呢？

Further, a new kind of manslaughter will appear, namely, failure-to-freeze. (As civilization continues its majestic advance, the categories of crime inevitably multiply.) Failure to get a body into a freezer, and failure to service a freezer, will probably count at least as negligent homicide.

In this connection, it is interesting to consider our attitude toward abortion, which is also a kind of cutting off of potential life. Abortion is a crime, but it isn't murder, and no funerals are held. Failure-to-freeze will not be taken as lightly, since the victim is more clearly a person, one who had a name and an identity and leaves a more definite sense of loss.

Freezing also offers an alternative to the abortion dilemma. If there are strong indications favoring abortion, but the people involved have strong feelings against it, possibly they might decide to remove the fetus by a careful operation and freeze it rather than destroy it, so that the potentiality of life remains.

Making freezing at death compulsory will at first be successfully opposed in the name of individual and religious freedom, somewhat analogously to the claims of certain Christian Scientists and snake-handling cults. But the courts have overridden the religious objections of parents to ensure proper medical care for dangerously ill children, and have allowed the police to interfere with the snake-handlers. Similarly, the relatives of the deceased will be compelled to freeze him.

還有，未來將會有一種新的殺人法出現，那就是沒有去進行冰凍。(當文明繼續其尊嚴的進步，罪行的種類，不可避免地一定會多元化。) 沒有將一個身體放入一個冰凍櫃，以及沒有去提供一個冰凍櫃，至少都將可能會被判定為是一種過失殺人罪。

　　和此有關聯的，去思考我們對墮胎的態度是頗具意義的，因為這也是屬於斬死潛在生命的一種。墮胎是一種罪行，但是它並非是一種謀殺，而且也不會為此而舉行葬儀。然而沒有去進行冰凍，就將不會被輕易放過，因為受害者明明就是一個人，一個有名有姓和有身份的人，而且會讓人有較為確實的失落感。

　　冰凍也提供給墮胎與否這個窘境一個替代方案。假使，有很強烈的指標因素都認為應該墮胎，但是當事者卻有強烈的感覺要反對的話，他們可能就可以決定利用細心的手術，來拿掉這個胚胎，並且將之冰凍，而不是將其完全地摧毀，這樣其生命的潛力就可以被保留。

　　把在死亡時的冰凍變成是強制的，這樣首先會被成功地反撲的，乃是以個人和宗教自由為名譽的，這是有點類似於某些基督徒科學家教派和弄蛇教派所主張的。但是法庭已經推翻這些宗教的反對，認為父母親應該要保證其病危的小孩，能夠獲得恰當的醫療照顧，而且容許警察去干預玩弄蛇的人。同樣地，去世者的親屬也將會被強制來冰凍他。

Suppose an adult of sound mind leaves explicit instructions that his remains not be frozen? This case will soon become more hypothetical than real. Before long nearly everyone will see the Golden Age shimmering enchantingly in the distance, and will not dream of relinquishing his ticket. Those that may remain stubbornly skeptical will realize they have nothing to lose: if by some chance they don't like what they see on awakening, they can then destroy themselves, or else climb back into the freezer. In practice, before long the objectors will include only a handful of eccentrics.

Widows, Widowers, and Multiple Marriages

In the Kingdom of Heaven, it is said, there is "neither marriage nor giving in marriage," and of course angels all love one another with indiscriminate determination, so that all the ex-wives and multiple husbands will simply sing in chorus. But on earth the resuscitees may have narrower views, and provision must be made for reunions which may not be entirely blissful.

A common form of the marriage vow says something about "until death do us part." If this be interpreted to mean permanent death, some brides and grooms will surely have second thoughts before promising to spend perhaps thousands of years with the same person. On the other hand, if temporary death is allowed to dissolve a marriage, as at present, and remarriages occur as usual, then many a widow will find herself, after resuscitation, facing two ex-husbands, of whom the less recent, the lover of her youth, is likely to be the dearer.

假若一個心智健全的成人，留下了清楚的指示，要求其遺體不被冰凍的話呢？這種情形不久將會變得較為是一種假想的，而非真實的。不會太久，幾乎每一個人都將會看到一個黃金的時代，迷人地閃爍在遠方，因而將不會想要去放棄他的入場券的。那些可能還是非常固執的懷疑者，他們也將會發覺其實他們無啥可失的：因為在他們醒過來時，假使看到某些他們不喜歡的，他們還可以再摧毀自己，或是再度爬回冰凍櫃中去。事實上不要多久，僅將會剩下一小綽古怪的反對者。

鰥寡以及多重婚姻

據傳說，在天上的國度中，"人也不娶也不嫁"，而且天使當然都是不分彼此地相親相愛，因此所有以前的老婆們和多個老公，都將會一起琴鳳合鳴。但是在地球上，復活的人可能將會有較狹隘的觀念，所以針對再度相聚一定要定出規章，因為一切並非全是幸福的。

在婚姻的誓言中，一個慣常的型式乃是去說些有關"直到死時我們才會分開"。假使這要被解讀成是指永久的死亡的話，哪麼一些新郎和新娘，在發誓要和同一個人相處或許會有好幾千年的時光之前，將一定會再度審慎地思考一下。另一方面，假使暫時的死亡，就被容許來解除一個婚約，就如同現在一樣，再度結婚可以照常進行的話，那麼有許多寡婦在其被復甦之後，將會要去面對兩個前夫，而其中較久之前的那一個，也就是她較年輕時的那個情人，就有可能會是較甜蜜的那一個。

In a few score years these questions may be meaningless. Who can be sure the institution of monogamy will persist? At present we are thoroughly committed to it, and yet one remembers wryly the moment in Shaw's Caesar and Cleopatra when a Briton expresses shock at a Roman custom. Caesar, speaking to another Roman, says: "Pardon him, Theodotus: he is a barbarian, and thinks that the customs of his tribe and island are the laws of nature." Just so; our tribal custom of monogamy is not a law of nature, and may eventually be replaced by . . . what? Perhaps group marriage, or no marriage at all, or marriage determined on a strictly individual basis by contract. With the biological functions and the nature of reproduction itself subject to scrutiny and deliberate change, no one can make a long-range guess with confidence.

A momentary digression here may be useful to point out that the religious notion of "natural law" is by no means so rigid a concept as many Catholic laymen, for instance, seem to believe. George W. Constable, writing in the Natural Law Forum of the Notre Dame Law School, has said: ". . . natural law consensus is not and cannot be static. . . If the conclusion of one qualified member of society is in conflict with the conclusion of another as to what the natural law is in any given case, then, ex definitione, each is justified in following his own lights. . . All are subject to correction, whether priest, king, or democrat." (13)

In the immediate future, some of the problems and their likely remedies are fairly clear.

在幾十年之內，這些問題都將會變成是無的放矢。誰能保證一夫一妻的體制將還會存在呢？在目前，我們乃是完全地在信守著它，在蕭伯納的凱撒大帝和埃及豔后克里歐佩特拉的作品中，我們都還不以為然地記得，當一個英國人對一個羅馬的風俗，表現出驚訝的那個時刻。凱撒對著另外一個羅馬人說，"寬恕他，提奧多塔斯：他是一個化外之人，而且他是認為他的部落和他的島嶼的風俗就是自然的定律。" 就像這樣，我們部落式一夫一妻制的風俗，並非就是自然的定律，而且最後都會被替代掉……是什麼呢？或許是集體的婚姻，或許是完全沒有婚姻，或許是完全是以個人為基礎，用契約來決定的婚姻。在生物的功能和繁殖本身的本質，都在接受重新的檢視以及審慎的變革之下，沒有人是可以信心十足地來做出一個長程的臆測的。

　　在此，一個暫時的出軌可能是有用的，它可用以來指出，宗教上所謂 "自然定律" 的觀念，絕對不會像是例如許多天主教門外漢所相信的那麼地死板。**George W. Constable** 在聖母法學院的自然法律論壇中寫過，他說："……對自然律法的共識，不是，也不可能是靜態的…………假使社會中的一個合法份子的論點，在有關所謂自然律法上的任何一個案例中，和另外一個人的論點是互相衝突的，那麼在超越定義之上，每一個人都有權利來依循自己的看法………而每一個人，不管是牧師、國王、或是民主人士，都還是要接受糾正。" **(13)**

　　在不久的未來，某些這類的問題和它們可能的解決辦法，將會是相當地清楚的。

The first marriage partner to die will leave demands on the survivor not formerly known, demands both emotional and financial. The freezee will want to awaken neither deserted nor impoverished, but to reclaim both his wife and his estate. The wife, on the other hand, may want to inherit everything, and may want to be free to console herself. What to do?

If we are talking about an average couple in the near future, so that the man dies at a moderately advanced age leaving a very modest estate, the result seems clear enough. The widow will be faithful. A decade or so of separation, at an advanced age, is not a high price to pay for emotional security. For the peace of mind of the first to die, this may even be formalized in law; under these circumstances, the widow of a freezee may be legally still married, and no more able to obtain a divorce than the wife of someone in an insane asylum.

Some may object that all this concern is unrealistic. After all, the resuscitees will not be the same people; they will be rejuvenated and overhauled, changed and improved (although not necessarily immediately) in physique and personality. The life will be new in a very drastic sense, and there may be no interest at all in the former spouse.

第一個將要死的結婚伴侶,將會對存活者留下一些前所未有過的要求,不僅會有在感情上的,也會有在財務上的要求。被冰凍者將會要求在其被復甦之後,不僅不會被遺棄,而且也不能是貧窮的,因而會要求來重新來獲得他的老婆和家產。但是在另一方面,其老婆可能會要求要能夠來承繼所有的東西,而且會要求能夠自由地來主控她自己。那這樣該怎麼辦呢?

　　假使我們現在所在談的,是有關在最近未來的一對一般的配偶,而其中的男人死在一個一般老的年紀,並且留下一筆普普通通的家產,那麼,其結局似乎就相當清楚了。這個寡婦將會是忠實的。因為在一個高齡之下,十幾年左右的分離,對感性上的安全感而言,所付出的代價應該不是有多高的。為了讓先死的那個人能夠心平氣和,這甚至還可能被型之以法律條文;在這種情境下,一個被冰凍者的鰥寡者,可能還是可以合法地再婚,但是就和某一個住在精神療養院的人的老婆一樣,不再可能去獲得一個離婚。

　　有些人可能會抗議說,這些顧慮都是不切實際的。畢竟,這些被復甦者,將不再被認為是同一個人了;他們將會被重新整修過,並且會被回春過,在身體上和個性上都會被改變和改善(雖然這不一定是馬上的)。整個生命將會是非常巨幅地更新,因而可能會對先前的配偶一點都沒有興趣。

The answer is that there must be a reasonable amount of continuity, or at the very least the anticipation of a reasonable amount of continuity (in personal relations), for otherwise the future would be too frightening altogether, and motivation would tend to evaporate.

Consider next a more difficult case, say where the survivor, even though aged, breathes a sigh of relief, thinking, "Good riddance to the bum! Thank heavens I don't have to put up with him any more." Or consider the case of a husband or wife dying in middle life, leaving dependent children. Notice I say, "consider," not "let us consider," because I have already considered them and find myself fresh out of answers. They will just have to be worked out - somehow.

Before leaving this topic, we might mention one possible solution to the problem of the young widow - one not put forward very seriously, but intended to remind the reader of the vast scope of the possibilities.

It is suggested by a news item relating that, in 1963, it is possible in Japan for a girl to go to a plastic surgeon, pay a fee of $50 to $100, and get herself a new maidenhead. (48) Her groom is thus spared the embarrassment of learning of her previous indiscretions. The next logical step, one presumes, is for the girl to go to a psychiatrist and have him hypnotically erase the memories associated with the original maidenhead!

其答案乃是，其中一定要有一個合理範圍內的連續性，或是至少能夠預期到一個合理範圍的連續性 (在個人的關係上)，因為不然的話，未來將會是太徹頭徹尾地可怕，而且人與人之間的善意，將會逐漸蒸發消失掉。

下面來思考一個較困難的案例，假使說，存活者雖然非常老邁了，但是想到說，"好不容易終於可以擺脫這個爛貨了！謝天謝地，我可以不用再忍受他了，"就鬆了一口氣。或是考慮到一個丈夫或是妻子死在中年，而留下了尚未獨立的小孩的案例。請注意我是說 "考慮到" 而非 "讓我們來考慮"，因為我已經思考過它們了，而且發現自己就是無法找到答案。但是無論如何，它們終究還是要被解決的。

在我們離開這個話題之前，我們還是要來提到年輕就鰥寡的問題的一個可能的解決方法，這問題雖然沒有被說得非常嚴重，但是還是將之提出來，以提醒讀者其中所可能涵蓋範疇的廣大。

有一則和此相關的新聞指出，在1963年的日本，一個女孩可能可以到一個美容醫生那裡，只要付美金50到100元的費用，就可以給自己一片新的處女膜。(48) 這樣就可以避免讓她的新郎知道她過去亂搞的窘境。這樣，我們就可以假想到下一個合乎邏輯的作法，就是這個女孩會去找一個心理醫師，要求他用催眠的方法，來洗掉所有和原來的處女膜有關的記憶！

Then she would be a maiden pure in every sense except that of history – and history, as everyone knows since H. Ford I, is bunk.

Our widow, then, makes the following arrangements. On revival, she lives with the second husband until they can separate by mutual consent - perhaps even until they are tired of each other. Only then is the first husband revived, and the wife meanwhile has her brain washed clean of the second husband by psychiatric or biopsychiatric techniques. Admittedly, the scheme in this simple form raises more problems than it solves, but it is only intended to be vaguely suggestive.

Cadavers as Citizens

Rumor has it that in certain political wards on Chicago's South Side, for example, it is possible by hallowed tradition for a recumbent corpse to be yet an upright citizen, since he retains his place on the roll of eligible voters. Perhaps, in some degree and sense, this custom will come to be fixed by law.

Two well established principles are involved: ". . . nor shall any State . . . deny to any person within its jurisdiction the equal protection of the laws" (U. S. Constitution) and "no taxation without representation" (Boston Tea Party et seq.). The frozen will be potentially alive; they will be property owners and tax payers. How must the law be modified for proper recognizance of these facts?

如此一來，除了在歷史之外，她在各個方面都完完全全是一個處女，而就如同大家都知道的，自從亨利福特一世以來，歷史就是無啥意義的。

　　因此，我們的寡婦將會進行籌措下列的事宜。在復甦之後，她會和第二任丈夫生活在一起，直到她們可以在雙方的同意之下分離－或許甚至會一直到她們相看兩相厭的時候。只有到此時，她才會將她的第一任丈夫復甦，而在同時，這個老婆已經利用精神醫學或是生物精神學的技術，將其頭腦中有關第二任丈夫的記憶清洗乾淨了。不必諱言地，這種過度簡化的說法所製造出來的問題，一定會比其所解決的問題還多，但是其用意本來就只是侷限在一個模糊的提示而已。

屍體的公民

　　譬如，據說在芝加哥南部的某些政治陣營中，根據他們神聖的傳統，一個已經在休眠的屍體，仍然可以是一個堂堂正正的公民，因為他還可以保留在合法投票者名單中的地位。或許在某些程度和意義上，這種風俗將可能藉著修法而讓其變成實質地合法。

　　其中牽扯到了兩條既存的法理：”...沒有任何一個政府...在其法律管轄的範圍之中可以來否定一個人的法律平等保護權”(美國憲法) 以及 “沒有代表權就沒有納稅義務”(波斯頓反茶稅示威以及其後續結果)。被冰凍者將可能會是活人；他們將也會是財產擁有人以及納稅人。法律應該要如何地修訂，方能夠對這些事實有恰當的認定呢？

At present, our voting laws are for the most part extremely simple – and simple-minded. One competent adult, one vote. Administratively, this is nice and tidy, but logically it is a ghastly mess. The whole area of voting rights and voting weights needs to be reexamined - not merely the question of lowering franchise age to eighteen, as has often been suggested, but the entire philosophy and rationale of representative government.

Should the vote of a man with four children count only as much as that of a father of two? The children are people, they have interests which can be furthered or damaged, and they are entitled to representation. Should a wellinformed voter swing only as much weight as the emptiest ignoramus? The very purpose of our republican government is to avoid this. Should not voting eligibility and voting weight depend on the specific issue and the degree to which the voter's interests are affected? It is already customary in certain areas for some issues to be voted upon only by property owners.

Perhaps another layer ought to be sandwiched in between the citizens and the legislatures. That is, any group of citizens might be permitted to delegate their votes to a chosen elector, who would be authorized to cast these votes in an election.

In any event, such an overhaul will surely, among other things, recognize the right of incompetents to certain kinds of representation.

在目前，我們大部分的投票法都是非常地單純－而且是有點天真的。一個具有行為能力的成年人，每人一票。從管理學上來講，這真的是乾淨俐落，但是從邏輯上來看，則會是一個天下大亂。有關投票權和每票權重，這整個領域是有需要被重新審視的－如已經時常被建議的，這不光是去將法定年齡下降到十八歲而已，而是有關政府代表性的整個哲學和倫理。

一個有四個小孩的男人的一票，是否僅應該被算成和一個僅有兩個小孩父親的一票一樣呢？小孩也是人，他們也是有一些權益，而此有可能被增益或是被損害，因此他們配獲得有其代表性的參與。一個知識淵博的投票人，難道僅應該和一個腦袋空白的笨蛋，一致地發揮同樣的權重嗎？我們的共和政府最最重要的目的，就是要去避開這個的。投票的資格和投票的權重，難道不應該依據個別特定的議題，以及投票者的權益受到影響的程度，而來決定的嗎？在某些議題中的某些領域，過去已經有過慣例，僅僅讓不動產擁有者來投票的。

或許在公民和立法委員之間，應該還要穿插入一個階層。那就是任何一個屬於公民的團體，可能可以被允許來將他們的票，委派給一個選舉出來的選舉人，他將被授權來在一個選舉中投下這些票。

在任何的情況下，類似這樣的整頓，除了有其它的功能之外，將一定可以肯定到這些沒有行為能力者的權利，而使其擁有某些種類的代表性。

Incompetents now form a small group, which is ignored in this respect; but the frozen will constitute an enormous body of influence which must be duly recognized and represented.

Potter's Freezer and Umbrellas

For failure to pay the premiums on one's freezer insurance, the death penalty seems a trifle severe. Hence society will be obliged to freeze the indigent. How fine, that the ne'er-do-wells will in the future escape both death and taxes! They will live on The Welfare and, dying, remain on The Welfare. To add insult to injury, on resuscitation they will be just as bright and shiny as the people who paid taxes. Is this justice? Ask me again in a thousand years.

For the further protection of the weak, the lazy, and the unlucky, the inheritance and bankruptcy laws will need working over. I shall not delve into this, except to remark that the quality of mercy may be displayed by ruling debts subject to simple interest only, while assets may accumulate compound interest.

Countless other legal problems remain to be first revealed and then handled. And while it is true that the freezer era will be the era of the Golden Rule, the fraternal outlook will become general only gradually, and even then there will be honest differences of interest and opinion.

沒有行為能力者，雖然現在是有組成一個個小團體，然而在這方面，卻還是被忽視的；但是在未來，當被冰凍者組成一個具有影響力的巨大團體時，這就一定會按道理地來被認定和被代表。

窮人的冰凍和保護傘

一個人為了沒有去繳納冰凍計劃保險的保險費而被判死刑，看起來似乎是有點小題大作。因此社會將應該有義務，來冰凍那些貧窮的人。這樣那些終生潦倒的人，在未來不僅將可以免去死亡，同時也不需要納稅，那真是太棒了！活著時，他們可以倚靠社會福利，死了，還是靠社會福利。以在傷口上灑鹽的觀點來說，這些人在被復甦之後，將會和那些有納稅的人一樣地光鮮亮麗。難道這樣是公平的嗎？請在一千年之後再問我這個問題。

為了要進一步能夠保護弱勢者、懶惰者和不幸者，有關遺產繼承和破產的法律，將需要被重新修訂過。除了認為或許可以藉著規定負債僅能用單利來累進，而資產可以用複利來累積，以便能展現出仁慈的格調之外，我是不想去深入探討這個問題。

還有不可勝數的法律問題，有待先被揭發，然後再將之一一處理。而且，雖然冰凍人的世紀的確會是一個黃金鼎盛的世代，但是四海皆兄弟的前景，卻只會逐漸地才會變成普及，而且甚至於到那個時候，還是會存在有一些真實的利益和意見上的差異。

For a considerable period we will have to bear in mind the immortal words of Ferguson Bowen: "The rain it raineth on the just / And also on the unjust fella / But chiefly on the just, because / The unjust steals the just's umbrella."

會有一段相當長的時間，我們必須要將 Ferguson Bowen 的不朽的話語深記在心頭："甘霖會滋潤好人／同時也會霑淋壞人／但是主要地還是好人，因為／壞人偷了好人的雨傘。"

CHAPTER VII

The Economics of Immortality

永生的經濟學

CHAPTER VII

The Economics of Immortality

Even though Professor John K. Galbraith and many others have described our society as "affluent," most people know in their bones this is balderdash. The fact that many people are much poorer does not make us rich. In 1958 the median American family income was only $5,050 (66), which might look good to a Hottentot but is scarcely tolerable by our own present standards, and which seems entirely intolerable if we dare lift our faces from the dust long enough to catch a glimpse of what may be and ought to be. Our wants - our realizable wants, in many cases pertaining to basic physical requirements of health and safety - greatly exceed our wealth.

This view - that our country by even modest standards is not rich but poor - is supported, for example, by Professor Edward C. Banfield of Harvard, who has written: "No one can possibly maintain that our economy is able to produce all of the goods and services that people want. We could not do that, or begin to do it, even if we all worked an 80 hour week . . . The fact is that much of our population is very poor.

第七章

永生的經濟學

雖然 John K. Galbraith 教授以及其他許多人，都把我們的社會描述成 "富足的"，但是大部分的人都深切地知道，這乃是一派胡言。事實上，使許多人變得較窮，並不見得會讓我們富有。在 1958 年時，美國一般家庭的平均收入僅有美金 5,050 元 (66)，這對蠻荒土人而言，看起來可能是挺好的，但是以我們自己目前的標準來看，則幾乎是不能忍受的，而且，如果我們有足夠的勇氣，從紅塵中抬頭長視，來看到我們可能是，而且應該是的情景，那麼一切現況看起來則是完全不能忍受的。我們的需要－我們可以實現的需要中，有許多是和健康以及保障上的一些基本實質的要求有關的－是大大地超過我們的財富的。

甚至於以一般的標準看來，我們的國家並非是富足的而是貧窮的這個觀點，乃是可以被例如像是哈佛大學Edward C. Banfield教授所佐證的，他曾寫道："沒有一個人可以來堅稱說，我們的經濟有辦法來生產出人們所想要的所有的東西和服務。就算我們每個星期工作80個小時，我們也是無法做到，或是開始來著手做它.......在我們的人口中，有很多人是非常貧窮的，這是一個事實。

In 1957, one of every seven families or unattached individuals earned less than $2,000, and the average of those who earned less was only $1,100. . . We are so far from suffering from Abundance that we cannot afford, for example, to rid our cities of slums and blight. A recent study . . . showed that to bring all our cities up to what professional planners consider an adequate standard would cost over $100 billion a year for 12 years. . . My own income is a comfortable one, but I wish it were 10 times what it is. I think I could make very good use of that much more. Most people, I expect, feel the same way." (4)

This being the case, one may tend to be daunted by the likely costs of a freezer program, direct and indirect, and still more by the prospect of immortality with its population problems. But on closer inspection these tough, new problems turn out to be not so new after all, and perhaps not so tough either.

Before coming to grips with specifics, we shall want to view the questions of wealth and population from Olympus. in preparation for this, it is most important to gain some appreciation of the infinite potential inherent in problem-solving machines.

The Solid Gold Computer

Everyone who reads the papers or watches TV knows by now that, whereas the first industrial revolution involved the replacement of human and animal muscle by machines, the second industrial revolution, now barely beginning, rests on the replacement of human brains by machines.

在 1957 年，七個家庭或是未婚的個人中，就有一個的收入是少於美金 2,000 元，而那些賺得較少的人的平均收入僅有美金 1,100 元........我們離會因為富庶而遭殃的情形還很遠，因為這是我們財力所不可及的，例如去拆除我們城市中的貧民區和老舊區。有一個研究顯示出......要將我們所有的城市，提昇到都市規劃專家們所認定的一個足夠的標準，這則要在 12 年中每年耗費掉美金一千億元.......我個人的收入已經是相當豐厚的了，但是我還會奢想著能夠賺上現在的十倍。我想我知道如何好好地來支配所多出來的錢。我相信大多數的人都會有同感。" (4)

既然如此，大家可能很容易地，就會被一個冰凍計劃可能要的花費所嚇退，不管其是直接的或是間接的花費，而且更為嚇人的，乃是永生的可能所可能帶來的人口問題。但是近一步審思這些，我們會發覺，這些新的問題畢竟不是那麼的新，而且可能也不算是新的。

在我們還沒有進入去拿捏特定的問題之前，我們想要先去從奧林帕斯的觀點，來審視財富和人口的問題。在準備這個的時候，先能去體會到在一些問題解決的機器中所蘊藏的無限潛力，乃是最重要的。

純金的電腦

每一個有在看電視或是讀報紙的人，現在都已經知道，第一次產業革命中所涵蓋的，乃是用機器來替代人的或是動物的體力，而現在剛剛開始不久的第二次產業革命，乃是架構在利用機器來替代人類的頭腦。

The computers already have remarkable problem-solving capacities, and it appears to be only a matter of time until they can "really think."

The invention of thinking machines, of automata with genuine intelligence, will of course have an importance difficult to exaggerate, quite aside from the prospect of immortality. This invention will obviously be in one sense the most important ever made, since it is equivalent to the invention of a magic lamp from which will stem other wonders without limit. There are many "philosophical" implications, some of which will be touched upon in later chapters, but at the moment our concern is with the economic impact.

Specifically, we want to lay the groundwork for the concept of unlimited productive and inventive capacity, through the agency of intelligent, selfreproducing and self-improving machines. The first aim is to convince the reader that such machines will indeed appear.

It is acknowledged in advance that anyone has the right, if he chooses, to reserve to humanity such words as "think," "imagine," "feel," and "live." When referring to machines, one may substitute the phrases "seem to think," "appear to display imagination," etc. With this understanding, then, the simpler terminology will be used in the discussion.

電腦現在已經具有令人驚嘆的問題解決能力，而且看起來似乎只要稍微假以時日，它們就將可能來 "真正地思想"。

屬於自動機器而且具有真正智慧的思想機器，其發明當然會有難以過度誇大的重要性，這和永生的期盼有著極大的不同之處。很顯然地，這個發明在某一方面，將會是前所未見最重要的，因為這就好像是發明了一盞阿拉丁神燈，從其中將會衍生出許多源源不絕的神奇。其中會有許多 "哲學上" 的意涵，在後面幾個章節中會談論到某一部份，但是當下我們所關心的乃是其在經濟上的衝擊。

特別是透過智慧的、會自我複製的和自我改良的機器的機能，我們將要來為無限生產力和創新發明能力的這個概念，來奠下基礎。然而，首要的目標乃是要先說服讀者，這種類的機器將一定會出現。

事先我們有認知到，任何一個人都有權來將例如 "思想"，"想像"，"感覺"，和 "生存" 這些字眼，全部回歸到人類的專屬，假使他選擇去如此想的話。而當一談及機器的話，我們大可以用 "看起來會思想"，"似乎呈現出想像力" 等字句來替代。有了這樣的理解之後，我們就可以用較為簡單的字彙來進行討論了。

Let us at once attempt to shatter the notion held by most laymen, and fostered by some scientists, that, while machines can calculate, they will never be able to show the higher qualities of thought, will never display originality, and will never transcend the limitations of their inventors. We shall first quote some expert opinion and then discuss some specifics.

Dr. J. L. Kelly, Jr. (Bell Telephone Laboratories) and Dr. O. G. Selfridge (Lincoln Laboratories, M.I.T.) say: "Now we believe that it is certainly logically and physically possible for a digital computer to do any sort of information processing that a man can. This includes thinking or invention, regardless of how broadly they are defined." (53) (One of these scientists is optimistic, the other pessimistic, about the length of time such developments might take, but this is of no great importance.)

Dr. Jerome B. Wiesner (former Special Assistant to the President for Science and Technology) has pointed out that machines may eventually rival the human mind in compactness of information storage and that they greatly exceed it in speed. Neurons cannot respond oftener than about 100 times per second, whereas electronic switching exceeds a rate of a billion per second. Nervous signals travel no faster than about 300 meters per second, whereas electric signals travel essentially with the speed of light, namely about 1,560,000,000,000,000 furlongs per fortnight, or a million times the nerve speed. From these and other considerations he concludes that ". . . one should ultimately be able to create thinking machines much brighter than the smartest human being, if presently unforeseen limitations . . . do not appear." (128)

讓我們立即試著來粉碎大多數外行人所持有的，以及某些科學家所鼓吹的一個觀念，那就是，雖然機器會進行計算，但是它們將永遠不可能展現出更高品質的思想，將永遠不可能展現出原創力，而且也將永遠不可能超越發明它們的人的極限。我們首先將引述一些專家的意見，然後再來討論某些特定的問題。

　　J. L. Kelly 二世博士 **(**貝爾電話實驗室**)** 和 **O. G. Selfridge** 博士 **(**麻省理工學院林肯實驗室**)** 表示，"現在我們相信，讓一台數位電腦來做一個人會做的任何種類的資訊處理，在邏輯上和實質上都是可能的。這還包括了思想和發明，不管它們的定義是如何地廣泛。" **(53)** **(**針對這種研發可能需要的時程，其中一個科學家是樂觀的，而另外一個則較為悲觀，但是這不是很重要。**)**

　　Jerome B. Wiesner 博士 **(**以前是美國總統的科技特別助理**)** 指出，總有一天，機器在資訊儲存的空間大小上，將勝過人類的頭腦，而且它們在處理速度上，也將大大領先人類。人類的神經元每秒鐘的反應不可能快過 **100** 次，然而電子的開關，每秒中則可以超過十億次的反應速率。神經元的訊號傳輸速度，每秒鐘不可能快過大約 **300** 公尺，然而電子訊號的傳輸速度，幾乎就等於光速，也就是每兩星期有大約 **1,560,000,000,000,000** furlong **(1/8** 英里**)**，或是神經傳遞速度的一百萬倍。從這些方面以及其它方面的考量，他總結說"………我們最後將可能創造出比全世界最聰明的人還要聰明許多的思想機器，假使現在我們所無法預知的一些限制……沒有出現的話。" **(128)**

Dr. Marcel J. E. Golay (Extraordinary Professor at the Technische Hogeschool, Eindhoven, The Netherlands) also believes that "mere size, complexity and speed may play the main part in transforming the 'stupid' computers of today into thinking machines which will teach us basically new concepts." (33)

A similar note is struck by Dr. W. Grey Walter, director of the Burden Neurological Institute, London, who believes that mere complexity may largely span the gulf between crude machines and sentient beings. {125}

Those who deprecate "mere" complexity forget that quantitative differences can mount up until they become qualitative differences. A very simple computer may only be able to add and subtract; but if we enlarge the computer sufficiently, although it is still only capable of addition and subtraction, it can now combine these operations in such diverse and complex ways that the result is multiplication and division, and even differentiation and integration, and more! A difference in degree may become a difference in kind.

Professor Norbert Wiener, the famous originator of cybernetics, believes that machines can and do transcend some of the limitations of their makers, and can be capable of originality. (101)

Dr. Marvin Minsky (Lincoln Laboratories, M.I.T.) says, "I believe . . . that we are on the threshold of an era that will be strongly influenced, and quite possibly dominated, by intelligent problem-solving machines." (74)

Marcel J. E. Golay 博士 (荷蘭 Eindhoven 科技大學的傑出教授) 也相信說 "光是在尺寸，複雜度和速度上，在將當今 "愚笨的" 電腦轉變成會思想的機器，就可能去扮演一些重要的角色，而基本上，這些機器將會教導我們一些新的觀念。" (33)

英國倫敦 Burden 神經學研究中心的主任 W. Grey Walter 博士，他也提出了類似的看法，他相信，光是在複雜度上的差異，就可能會造就了粗糙機器和有感知物種間的大鴻溝。(125)

那些抨擊 "光是" 複雜度的人都忘記了，數量上的差異可能會逐漸累積，直到它們變成質量上的差異。一台非常簡易的電腦可能只會做加減；但是如果我們將此電腦變得夠大的話，雖然它還是只會做加減，但是它現在就可以將這些運算，以各種不同的和複雜的方法混合，而其結果就等於是乘和除，甚至於是微分和積分，甚至於更多！程度上的差異，可能就會變成一個種類上的差異。

神經機械學有名的創始者 Norbert Wiener 教授相信，機械的確可能也會超越它們的創造者的一些極限，而且也可能會具有原創能力。(101)

Marvin Minsky 博士 (麻省理工學院林肯實驗室) 表示," 我相信.......我們目前正處於一個世紀的門檻點,此世紀將會強烈地被智慧型的問題解決機器所影響,而且頗有可能會被其所操控。" (74)

The list of optimists could be extended indefinitely. Looking for pessimists, there seem to be very few among experts actually working in the field. A semi-skeptic is Dr. Mortimer Taube, who has devoted a whole hook (114) to scolding those scientists who (a) are "over-optimistic" regarding rapidity of progress, (b) exaggerate the closeness of analogy between brain and computer, and (c) assume a materialistic, mechanistic universe and a lack of fundamental distinction between man and machine. Perhaps his nagging injects a healthy note of caution, especially with respect to time tables. But (b) and (c) need not worry us; we may not care very much what methods the machines employ, or whether they have "really" any awareness. Dr. Taube does not place any limits on the objective capabilities of the machines.

Looking now at some actual accomplishments to date, we note that Dr. Arthur Samuel (I.B.M. scientist) is reported to have designed a checker-playing machine which regularly beats him at checkers. (101) Already we see a machine which in one narrow way transcends the intellectual powers of its maker. It is true, as we are so often reminded, that this machine can only do what its program tells it to do, and that the programmer could do the same thing himself (more slowly) if he wished. But while its moves are predictable in principle, in practice they are unexpected.

Dr. S. Corn has discussed some of the ways to endow a machine with learning ability. (34) And it is well known that machines can be programmed to learn very easily, if elegance is no object.

樂觀者的名單可以說是不可勝數。現在讓我們來找尋一些悲觀者，但是看起來，事實上在此領域鑽研的專家中，似乎很少是悲觀的。有一個半懷疑者，那就是 **Mortimer Taube** 博士，他有一整本書 **(114)** 專門來斥責一些科學家，他們是 **(a)** 對有關進步的速度 "過度樂觀的"，**(b)** 誇張了人腦和電腦之間類比的接近度，以及 **(c)** 把宇宙假設成物質的和機械的，而缺乏對人和機器之間差異的基本認知。或許他的嘮叨可以注入一個健康性的警語，特別是在有關時程方面的。但是我們不需要去擔心 **(b)** 和 **(c)**；我們對這些機器到底用什麼方法，或是它們是否 "真的" 有任何的感知，可能是不需要去那麼地在意。**Taube** 博士針對這些機器客觀上的能力，其實並沒有下定任何的限制。

　　現在讓我們來看看一些到目前為止的成就，我們注意到了 **Arthur Samuel (I.B.M.的科學家)** 的報導，他已經設計出一台會下西洋棋，而常常會打敗他的的機器。**(101)** 我們可以看到一台機器，在某一狹小的範圍內，已經超越了它的創造者的智能。的確，就如同我們常常被提醒的，也就是這台機器僅能執行它的程式命令它去做的，假使程式設計師想要的話，他自己也可以做同樣的事情 (只是較慢而已)。雖然在原理上，它的舉動是可以預測到的，但是在事實上，它們都是突發之舉的。

　　S. Corn博士有討論過某些方法，來將學習的能力賦予一台機器。**(34)** 而且，假使優雅不是重點的話，那麼機器可以被程式化來快易地學習，這乃是眾所皆知的。

For example, a machine with a large enough memory could be easily programmed to learn chess. It would start out playing poorly, but would not repeat mistakes, so its game would slowly improve. If it played enough games against the best players, it would eventually surpass all of them. (In fact, it could even learn by playing against itself.) Many ways are being studied to improve economy, elegance, or subtlety.

Dr. Herbert A. Simon and Dr. Allen Newell (Carnegie Institute of Technology and The Rand Corporation) have described other recent computer achievements:

"[There is a program that can] discover proofs for mathematical theorems-not to verify proofs, it should be noted, for a simple algorithm [procedure] could be devised for that, but to perform the 'creative' and 'intuitive' activities of a scientist seeking the proof of a theorem.

"At least one computer now designs small standard electric motors (from customer specifications to the final design) for a manufacturing concern.

"The ILLIAC, at the University of Illinois, composes music and I am told by a competent judge that the resulting product is aesthetically interesting." (106)

譬如一台機器，如果具有足夠大的記憶體，就可以很容易地被程式化來學習下西洋棋。它剛開始可能會下得很菜，但是因為它是不會去重複錯誤的，所以它的技藝就會緩慢地精進。假使它和一些最好的棋士對奕過足夠的次數，它最後一定會超越所有的人。(事實上，它甚至於還可以藉著和自己的對奕而來學習。) 目前還在進行一些研究，是在改進其經濟性，優雅度，或是細緻度。

Herbert A. Simon 博士和 Allen Newell 博士 (卡內基科技學院和 Rand 公司) 有敘述過其它新近在電腦上的研發成果：

"[有一種程式可以來] 找出一些數學理論中的證明-雖然這並沒有去證實這些證明，我們應該注意，我們是可以設計出一個簡單的演算 [程序] 來執行它，但是它卻是沒有辦法去進行，一個科學家在追尋一個理論的證明中，所需要的'創造性'和 '直覺性' 的一些活動。"

"至少現在已經有一台電腦，可以為了一個製程上的構想，設計出小型的標準電動馬達 (從客戶的規格一直到最後的設計)。"

"在伊利諾大學的ILLIAC電腦會編作音樂，而且有一個蠻不錯的法官告訴我，其所編作出來的樂曲竟然是那麼地優美饒趣。" (106)

Let us now turn to evidence that machines can exhibit life-like behavior, including reproduction, "purposive" activity, and homeostasis (maintenance of internal conditions within permissible limits, in spite of changes in the outer environment).

The latter two are exemplified, in a crude and elementary way, by the "mechanical tortoises" of Grey Walter. (125) These are little electrical-mechanical devices which propel themselves on wheels, and wander around, in a manner suggestive of "curiosity," until their batteries get low; then they seek an electrical outlet and plug themselves in for a recharge. In seeking the outlet, one will look for ways around obstacles, and will probe and try in unpredictable ways until it either succeeds or "dies." This is not a bad imitation of one of the main features of life, say on the level of microorganisms.

Professor Kemeny has discussed a "reproducing" or self-duplicating machine proposed by von Neumann. This is a device which is extremely simple compared to any biological organism, with a "body" of about 32,000 simple parts and a "tail" of about 150,000 simple information units, analogous to the units of heredity in a living plant or animal. The tail serves as a blueprint describing the machine. In a suitable environment, the machine can make a copy of itself by reading the blueprint in the tail; after making a daughter machine, it also copies the tail and attaches the new tail to the daughter, which is then in business for itself. (Sex plays no part, and the daughter is exactly like the mother, except for possible "mutations" by accidental interference or malfunction.) (54)

現在讓我們把話題轉到機器可以展現出類似生命行為的證據上，其中也包括了複製能力，"有目的性" 的行為，以及內在的平衡力 (不管外在環境的各種變化，在容許的限度內，可以維持其內在的各種狀況)。

後面兩個行為，可以藉著 Grey Walter 的 "機械烏龜" ，來作一個粗糙和基礎式的舉例說明。(125) 它們乃是一些小型的電機機械，用輪子來驅動它們自己前進，而且以一種好像是 "很好奇的" 樣子，到處亂竄，直到它們的電池快沒有電為止；然後，它們就會去尋找一個電源插座，並且將自己插入其中，以便能夠充電。在尋找插座的過程中，每一隻都要在障礙物之外，去找出一些路徑，而且要以一種無法預期的方式來探索和嘗試，直到它不是成功就是 "成仁" 方休。如果以微生物的層次來看的話，這似乎是對生命的一個主要特質的一個不遜的模仿。

Kemeny 教授有討論過由 von Neumann 所提議的一種 "會繁殖的" 或是會自我複製的機器。這個機器和任何生物性有機體比起來可說是非常地簡單，它具有一個 "身體" ，其中大約有 32,000 個簡單的零件，而且有一根 "尾巴" ，其中大約有 150,000 個簡單的資訊單元，這有一點類似植物或是動物中的遺傳單元。這根尾巴乃是充當描述這台機器的一張藍圖。在恰當的環境中，這台機器可以藉著讀取尾巴中的藍圖，來製造出一個自己的拷貝；在製造出一個子代機器後，它也會拷貝出尾巴，然後再將新尾巴附著在其子代，之後子代就要去自力更生了。 (此中性並沒有扮演任何的角色，而子代是完全和其母親一樣，除非有因著意外的干擾或是錯誤所可能造成的 "突變"。) (54)

British geneticist L. S. Penrose has also described self-reproducing machines. (89) He has designed mechanical models to have many analogies to chemical and biological properties of living things. The machines possess only a few parts, analogous to molecules in living matter. A mechanical scheme governs logic and programming, controlling correct assembling of parts. The scheme uses only hooks and latches, depending on gravity for their action. Parts are arranged at random on a flat surface, which vibrates to provide the necessary energy, producing motions analogous to the thermal agitation of molecules in nature. Each part has different states or conditions, with different potential energies. If a complete machine (called a "seed") is present, it causes the parts to rearrange themselves into copies of the first machine; if the seed is absent, there is no "spontaneous generation." In some models, a seed can contain indefinitely long chains of information-storing units, which can be likened to the chains of molecules in the chromosomes of living organisms. Some models have actually been built and successfully operated. (54)

It is true that von Neumann's machines and Penrose's machines are at once too simple and too dependent on special environments to have more than theoretical interest, although this theoretical interest of course is extremely important. But Dr. Edward F. Moore (Bell Telephone Laboratories) thinks that in only ten to fifteen years, with an effort that might cost as much as a half billion dollars, economically useful self-reproducing machines might be made.

英國的基因學家 L. S. Penrose，他也曾經描述過一些會自我複製的機器。(89) 他有設計出一些機械的模型，具有許多類似生物的化學和生物學上的特性。這些機器僅擁有少數的零件，有點類似在生命物質中的分子。它們具有一個機械性的系統，來掌管邏輯和程式，以控制各部件的正確組合。這個系統所利用的僅是一些鉤子和閂鎖，而且是靠著地心引力，來進行它們的作用。這些零件是被隨機地擺放在一個平面上，藉著震盪此平面來提供所需的能源，而來製造出類似於自然界中分子熱力激盪的一些動作。每一個零件因著在位能上的差異，都處於不同的能階或是狀態。假使有一台完整的機器 (在此稱為一個 "種子") 存在，它就會使得這些零件重新排列它們自己，而組成一些如第一台機器的拷貝；假使種子不在的話，那麼就不會有 "自發性的產生"。在某些模型當中，一個種子可以包含有許多長度不定的資訊儲存單元的長鏈，這些可以被比擬成生物染色體中的分子鏈。有一些模型也已經實質地被建造出來，而且已經成功地在運轉了。(54)

von Neumann的機器和Penrose的機器，的確一時看來有點太過於簡單，而且太過於倚賴一些特別的環境，以至於都無法跨越出理論上的興趣，當然這個理論上的興趣也是非常地重要。但是Edward F. Moore博士 (貝爾電話實驗室) 認為，在十到十五年之內，可能只要花上五億美元代價的一個功夫，就有可能製造出一些符合經濟效益的會自我複製的機器。

These would be self-contained sea-going mining or harvesting machines, which would bring back minerals or processed ocean crops. While on the job, they would power themselves by sunlight or by the food and fuel they find, and they would also build others of their kind. When they had produced enough new machines and collected enough of a harvest, they would swim dutifully home. They would be mechanical slaves which would enrich us not only by working, but also by breeding. (75)

For most uses, it will not be necessary for machines literally to reproduce themselves. They will, however, be required to design new and smarter machines, or to design improvements in themselves. And computers have already been used to assist in designing new computers. (101) The implications are obvious, and stupendous.

And now at last, having taken this long but interesting detour, we are ready to climb Mt. Olympus and appreciate the view.

The View from Olympus: How Rich Can We Get?

If we only assume progress continues more or less as it has done in this century, we shall grow richer rather rapidly. In 1958 the median U.S. income for a "consumer unit" (family or unattached individual) was $5,050. (66) Since about 1890 the yearly increase in productivity per capita has averaged around 2.3 per cent. (28)

這將可能是一些能自給自足的深海採礦或是捕集的機器，它們可以帶回來礦物或是加工海洋作物。在出任務的時候，它們會利用陽光，或是它們能夠找到的食物和燃料，來給自己能量，而且他們也會製造出和它們一樣的其它機器。當它們已經製造出足夠的新機器，而且也採集到夠多的作物後，它們就會乖乖地游行回府。它們就是一些機器的傭人，不僅經由工作，同時也藉著繁殖來讓我們更為富足。(75)

　　在大多數的應用上，其實並非真的一定要讓那些機器去繁殖它們自己。但是它們將會被要求來設計出一些較新較聰明的機器，或是來在其本身之上，設計出它們自己的改良型。其實有一些電腦已經被用來在輔助一些較新型電腦的設計了。(101)這裡面的意涵雖然看是淺顯的，但是卻是廣大的。

　　在歷經了這個雖然漫長，但是卻是有趣的繞道，現在我們終於準備要來攀登奧林帕斯山，並且要來欣賞其景觀了。

奧林帕斯的景觀：我們能多富有？

　　就算我們假設進步的速度大約僅是和這個世紀的進展一樣，我們也都將頗快地就會變得越來越富有。在1958年，美國一個 ”消費單元” (一個家庭或是未婚的個人) 的年收入中值是5,050美元。 (66) 大約自從1890年開始，每一人每年生產力的增加率，平均大約有百分之2.3。(28)

If we assume an average rate of increase of income of 2.5 per cent annually for the next 300 years, and if we assume no inflation or other disturbing influence (remembering that the statistics we used referred to real productivity and not flexible-dollar productivity) then in the year 2258 the median income will be over $8,000,000 a year! This is no fantasy, but a conservative projection; you will actually receive that much money every year, in terms of today's prices. The average woman then will have much more spending money than any movie star does today - and still more important, will have much more to spend it on.

(It may be objected that this picture is oversimplified, since, for example, it neglects the questions of relative land prices and of taxes. But unless there is a monopolistic landlord class, and unless the taxes are wasted, these considerations will make little difference.)

In any case, all this is merely preliminary. If we take a really long view, if we strip away all the nonessentials and disregard all the immediate problems, the production of wealth depends simply on the availability of matter, energy, and organization.

The kind of matter doesn't matter: with the right techniques and enough energy, any kind of atom can be transmuted to any other kind, in principle if not yet in practice; and the right kinds of molecules and higher complexes can be produced or reproduced. From our seat on Olympus, these are mere details.

假使我們假設在未來的 300 年中，每年平均收入有一個百分之 2.5 的增加速率，並且假使我們假設其間沒有通貨膨脹，或是其其它的擾亂影響 (請切記我們所用的統計法中，乃是指實質的生產力，而不是指會變動的貨幣生產力)，那麼到 2258 年，每年的收入中值將會超過 8,000,000 美元！這並非僅是一個幻想，而是一個保守的預測；你真的每年將會得到這麼多以今天的物價為基準的錢。一般的女人，將會有比現在任何一個電影明星還要多許多的錢可以花費——而且更重要地，將會有更多的東西可以買。

　　(可能一定有人會反對，認為這種景象乃是過度簡化的，因為例如這忽略了土地相對價格和稅賦上的一些問題。然而，除非其間會出現一個地主階級的壟斷，而且除非所有的稅賦都被浪費掉了，否則這些顧慮都將不會造成任何影響。)

　　不管如何，這一切都還僅是初步的觀點。假使我們對其觀察真的夠深入，假使我們能夠將所有不重要的因素去蕪存菁，並且能夠先忽視所有當下的問題，我們將會發現財富的產生，只是單純地和可供利用的物質、能量和組織有關而已。

　　物質的種類並不重要：只要有恰當的技術以及足夠的能源，假使在現實上還不行，至少在理論上任何種類的原子，都可以被轉變成另外一種；而且所要的那類的分子以及更高的複合物，都可以被生產和複製出來。從我們在奧林帕斯的制高點來看，這些都僅是細節上的事務而已。

Matter, of course, is in practically inexhaustible supply in the earth, the planets and satellites, and, if need he, the sun and even other star systems.

Energy will also be available virtually without limit. Nuclear fission energy is becoming cheaper, and John E. Ullman of Columbia University has predicted that by 1968 it will become as cheap as energy from conventional sources, and rapidly thereafter much cheaper. (122) It is well known that all our foreseeable needs for many centuries could be met (at prices not yet competitive) either from the sunshine reaching the earth's surface or from fission of low concentration uranium in granite. When the fusion problem (controlled thermonuclear reaction) is solved, there will be another nearly boundless supply of fuel in the deuterium of seawater. There is also the possibility, if a combination of circumstances should make it useful as a stop-gap, of setting up solar power stations on Mercury, where there is no atmosphere, a permanent day side, and a radiation intensity more than six times that on earth.

Our trump card, finally, is that unlimited organizing capacity is also in sight, in the shape of intelligent, self-propagating machines. Such a machine need only show a small profit: that is, it must be able to reproduce itself from scratch and also do some directly useful work before it wears out. This is enough to ensure, on the compound-interest principle, that starting with only one machine we can in sufficient time have as many machines and as much wealth as we please. One expects, of course, that in practice the profit margin will be ample and the machines can produce any desired amount of wealth with little time lag.

在地球、行星和其衛星中，實質上，物質當然是取之不盡用之不竭的，而且如果需要的話，太陽，甚至於其它星系也還有。

　　能源也將會是無限量地垂手可得。核子分裂的能源越來越便宜，而哥倫比亞大學的 **John E. Ullman** 預測說，到 **1968** 年，它就會和傳統的能源一樣地便宜，而且之後很快地還會變得越來越便宜。 (122) 大家都知道，從照射到地球表面的陽光，或是從裂解花崗岩中低濃度的鈾 (以目前尚未具有競爭力的價位) ，我們未來好幾百年中可以預見到的能源需球，就可能可以獲得滿足。而當核子融合的問題 (熱核能反應被控制) 獲得解決後，在海水中的重氫，將會是另外一種幾乎是取之不盡用之不竭的能源。另外還有一種在水星上面設置太陽能發電廠的可能性，假使各種狀況條件的結合得宜，使其變成是一種有用的權宜之計的話，因為在那裡沒有大氣層，有永久白晝的一邊，而且其光照射強度是超過地球的六倍以上。

　　最後，我們還有一張的王牌，就是以智慧、自我繁殖型式現身的機器中，其威力無窮的組合能力也是指日可見了。這一種機器僅需要能呈現出一點點的益處就夠了：那就是，它必須要能夠從零開始複製它自己，而且在它掛掉之前，也要能夠直接地執行一些有用的工作。基於複利的原理，這樣只要以一台機器來開始，在假以足夠長的時間，就可以確保我們能夠來擁有我們所想要的機器數量和財富。當然在事實上，我們預期這個的利潤空間是巨大的，而且這些機器不需耗時多久，就可以產生出任何我們想要的財富數量。

In a simplified, representational sense, then, one may picture the Golden Age society in which every citizen owns a tremendous, intelligent machine which will scoop up earth, or air, or water, and spew forth whatever is desired in any required amounts - whether caviar, gold bricks, hernia operations, psychiatric advice, impressionist paintings, space ships, or pastel mink toilet rolls. It will keep itself in repair, and in fact continuously improve itself, and will build others like itself whenever required by an increase in the owner's family.

It is clear that in the long run, as long as the machines reproduce themselves faster than the people, there can be no economic problem - unless we run short of space. Let us next size up the ogre of "population explosion."

The View from Olympus: How Fast Can We Spawn?

First of all, we must recognize that population problems and all the attendant difficulties will inevitably arise, with or without a freezer program. The freezers may exacerbate these problems, but will not create them. With or without freezers, there will soon be increased longevity. With or without freezers, there will eventually be an indefinite life span. Since solutions must be found anyhow, we may as well make them good enough so that our own generation and those immediately succeeding can share the Golden Age.

In fact, the population problem already exists, without freezers and even without extended longevity, simply as a result of natural increase. In many parts of the world it constitutes a serious economic and political problem right now.

以簡化而具有代表性的話來講，我們可以將黃金盛世的社會，描繪成其中每一個公民都擁有一台美妙聰明的機器，它將會伸入土壤，或是空氣，或是水中吸取原料，而來吐出所需要數量的任何想要的東西——不管是魚子醬、金塊、疝氣手術、精神諮詢、印象派畫作、太空船、或是柔軟貂皮質的衛生紙。它將會把自己保養的好好的，而且事實上，會隨時增進改良自己，並且只要當主人的家庭中有新增的需求，就會製造出其它和自己一樣的機器。

很明顯地，長久下來，只要機器複製自己的速度快過人口增長的速度，那就不會有經濟上的問題－除非我們的生存空間不夠了。下面讓我們來對"人口爆炸"這個怪物做一個評審。

奧林帕斯的景觀：我們能繁衍多快？

首先我們必須認知到，不管有沒有冰凍人的計劃，人口問題和其尾隨的一些困境，不可避免地一定會發生。這些冰凍人雖然會讓這些問題更為惡化，但是絕對不會是其罪魁禍首。不管有沒有冰凍人，人類壽命很快也會被增長。不管有沒有冰凍人，壽命長度最後也會變成無限長。既然不管如何，一定會找到解決方案，我們不如將它們搞好一點，好讓我們自己這一代以及那些馬上要承繼我們的人，都能夠享受到這個黃金盛世。

事實上，就算沒有冰凍人，甚至於沒有壽命的延長，光是自然的成長，人口的問題早已存在了。在世界許多地方，它現在就在造成一個嚴重的經濟上和政治上的問題。

In the United States the population increased from 132 million in 1940 to 151 million in 1959 and 179 million in 1960, with an estimated total of 375 million in the year 2,000, based on moderate assumptions. (8) But the Malthusian doctrine of population always outrunning food supply has here long been proven false: the record shows that our birth rate is responsive to economic conditions and the general outlook. (8) Similar remarks apply to Europe.

The prospect in other parts of the world might seem grimmer. China had about 654 million people in 1960, and if present trends continue will have 894 million in 1975. (8) But the government of China, despite its greed for cannon fodder, seems to have realized the folly of unrestrained growth and is promoting birth control, according to many reports. And the Japanese, once extremely fecund, have exercised their admirable intelligence and cut their birth rate to about that of the United States. (8)

In India, the population has grown from 361 million in 1951 to 461 million in 1963 an increase of 100 million in twelve years! But the government sponsored birth control program is reported to be making headway, with the birth rate in Bombay down from forty per year per thousand to twenty-seven, and with the beginnings of success in the countryside. (19)

There is some evidence that Roman Catholic opposition to birth control will recede. A prominent Catholic gynecologist, Dr. John Rock, has received wide publicity for his views favoring birth control.

在美國，人口從 1940 年的一億三千兩百萬增加到 1959 年的一億五千一百萬，到 1960 年的一億七千九百萬，而根據一些保守的假設，到 2000 年，預期總共會有三億七千五百萬的人口。(8) 然而馬爾薩斯認為，人口的膨脹一定會超過食物的供給的原則，早已經被證明是謬誤的：過去的紀錄顯示出，我們的出生率是會依著經濟狀況和一般局勢而有所變化的。(8) 同樣的結論也可以加諸在歐洲。

世界其它地方的前景看起來可能就較為淒慘點。中國在 1960 年時，大約有六億五千四百萬的人口，而如果以目前成長的趨勢持續下去的話，到 1975 年，將會有八億九千四百萬。(8) 中國政府雖然對槍砲彈藥還是貪得無厭，但是似乎也已經警覺到其對人口成長不設限的愚蠢，根據許多的報導，中國也正在大力推廣生育控制了。而曾經是非常多產的日本人，也已經運用了他們可佩的智慧，將其出生率減少到大約和美國的一樣。(8)

在印度，人口從 1951 年的三億六千一百萬增加到 1963 年的四億六千一百萬，在十二年間就有一億人的增加！但是其政府所主辦的生育控制計劃，據報導是有在獲得長足的進展，在孟買，出生率已經從一年千分之四十降到千分之二十七，而在鄉下地方也開始有所進步。(19)

有某些證據顯示出，羅馬天主教對生育控制的反對，將會逐漸地式微。一個著名的天主教婦產科醫生John Rock博士，他因著贊同生育控制的觀點，而受到大肆的報導。

In 1963 he is reported as saying, "The Catholic Church is in no way an obstructive agent to what is good for humans. The church does not sidestep responsibility . . . Not all of the church is done up in red petticoats and Roman collars. A large part is the lay church, which does not intend to be misled in obstructing its own welfare." Of those lay members, he said that 95 per cent of those who have expressed an opinion on his birth control plan have given their approval. (92)

All in all, there is good reason to believe that population will not run far beyond desirable limits, although some countries, especially in Africa and Latin America, may lag in progress. Human stupidity is formidable but not invincible.

It should also be noted that the freezer program itself will help speed the adoption of a reasonable birth control program, and perhaps of a general eugenic program. The long view will tend to make everybody more foresighted and aware of responsibility in all areas, including this one.

Granted that population can be controlled and that the actual course of events will be the sensible course, what is the freezer population likely to be when the Penultimate Trump is sounded?

One might guess that everyone would be satisfied with two children at about age thirty. Having fewer might tend to annihilate the family; having them earlier might build the frozen population too rapidly. Increasing the average child-bearing age to thirty would reduce the population for a while, but it would then stabilize.

在 1963 年，據報導他表示說，"對人類有好處的事情，天主教教會絕對不應該是一個阻擾的因子。教會並沒有逃避其責任........並非所有的教會都會是完全盛裝著裙袍和羅馬領子的。大多數的教會乃是世俗教會，他們不會想要被誤導，而來斷送其自己的福祉。" 在這些世俗的成員中，他說有百分之 95 都有表示對他的生育控制計劃給予贊同的意見。(92)

雖然某些國家，尤其是在非洲和拉丁美洲的國家，其進步可能較為落後，但是畢竟有很好的理由可以相信，人口將不會逃脫到遠超過可以接受的限度。人類的愚蠢雖然可怕，但是並非是無法克服的。

還應該強調的就是，冰凍人計劃本身，將有助於加快採取合理的生育控制計劃的腳步，而且可能是一個普遍優生的計劃。由長遠的觀點看來，每一個人都是趨向於更具有前瞻性，而且會知道他在各方面應負的責任，當然也包括這一方面。

如果人口可以被控制，而且整個事情的達成的實際路徑，都將是合情合理的，那麼當倒數計時的號角響起之時，冰凍人的人口將可能會有多少呢？

我們可以猜想，每一個人都會滿足在大約三十歲時生育兩個小孩。如果再生少一點，這可能就會斷掉家庭的煙火；假使有的較早有小孩的話，這就可能會過快地累積被冰凍的人口。將養育小孩的平均年齡提高到三十歲，在短時間內，人口將會減少，但是之後就會穩定下來。

If we consider the whole world, with a base population of, say, four billion, then the frozen population would increase by four billion every thirty years. If it takes 300 years for civilization to reach the immortality level, there would then be some forty billion people to revive and relocate - if we assume, for simplicity, that it all happens at once. The figure of 300 years is more or less picked out of a hat, of course, since we have no clear idea either of the extent of the problem or of our rate of progress; but the outlook with respect to thinking machines is so encouraging, and the rate of progress will be so steeply exponential once thinking machines exist, that it is difficult to suppose that any problem we are now capable of posing could take much longer.

There is ample room on our planet for forty billion people. Most of the land surface is thinly populated, with vast areas of the antarctic, the arctic, the jungles of South America and Africa, and the deserts of Australia, Asia, Africa, and the United States virtually empty, waiting to be made habitable and productive.

Agricultural and industrial techniques already known or in early prospect can probably handle a population of fifty billion, according to Professor Richard L. Meier of the University of Chicago. (67) Hence conditions at the opening of the era of immortality, based on our assumptions, would not be too bad, even without unforeseen breakthroughs - but what about the long ages following?

Retaining our seat on Olympus and assuming all problems of dissension will be solved and a reasonable course navigated, there seems very little cause for concern. First of all, if no other solution were in sight for a certain period of history, the people could simply agree to share the available space in shifts, going into suspended animation from time to time to make room for others.

假使我們將整個世界納入思考。譬如說以一個四十億的人口基礎來看，那麼每經過三十年，被冰凍人口就會增加四十億。假使文明要達到永生的境界需要 300 年，那麼到時大約就會有四百億的人要被復活和安置─為了簡化，假設說這一切都發生在同一時間點。當然這個 300 年乃是隨興瞎掰的，因為我們對這個問題的廣度，或是我們對進步的速度，都還是不夠清楚；但是有關會思想機器的前景，就非常地樂觀，只要會思想的機器一出現，進步的速度就會變成一條非常陡峭的指數曲線，所以我們當前有可能想到的問題，很難假設說會耗時多久方得以解決。

　　我們的地球是有足夠的空間來容納四十億的人口。地球大部分的土地表面，人口都非常地稀疏，還有南極和北極的廣浩地區，南美洲和非洲的叢林，以及澳洲，亞洲，非洲和美國的沙漠，實質上都還是渺無人煙，而等待著我們將其變成可居住和可生產的。

　　依據芝加哥大學 Richard L. Meier 教授，以已知的和在開發初期的農業和產業技術，就可以供給一個五十億的人口。(67) 因此根據我們的設想，在永生紀元的開端時，其狀況可能不會太糟糕，就算說沒有預期中的技術突破─然而，在其接續下來的長久年代之後會如何呢？

　　我們只要坐穩奧林帕斯的寶座，並相信所有紛擾的問題都將會解決，而且只要航行在一條理性的路徑上，一切就沒有理由好擔憂了。首先，假使在某一歷史時段中，看不見其它的解決方案，那麼人們只要同意輪流分享可用的空間，偶而進入活體休眠來讓出空間給其他的人就行了。

But the main point is that we can regard the available space as unlimited, remembering that we will sooner or later have unlimited wealth. For example, we could honeycomb the earth to a great depth, multiplying the usable surface. We could colonize other planets and satellites of the solar system, if appropriate at a certain stage in history. Beyond that, when our machines become numerous enough and big enough and small enough, we can simply use the mass of other planets, and even mass from the sun, actually to create thousands of new planets just like earth! Nobody would have to live underground.

Beyond that still, if we choose to breed fast enough and long enough to make it necessary, we can go to the stars. Strange and even wild as these possibilities may seem, they are nothing more than simple consequences of the concept of unlimited productive capacity, which in turn is a simple consequence of the concept of self-propagating, intelligent machines.

In the long run, then, neither costs nor population pressure need worry us. But now it is time to come down off Olympus and consider some of the very real and possibly dangerous intermediate problems.

The Cost of Commercial Freezers

One might expect a freezer program to multiply mortuary costs by a sizable factor. However, let us investigate this question a little.

但是重點乃是，我們可以將可用的空間認定為是無限的，請務必記得，遲早我們將會擁有無限的財富的。例如，我們可以將地球更深度地蜂窩化，來使得可用的表面積加倍。假使在歷史的某一個階段，時機恰當的話，我們還可以殖民到其他太陽系中的行星和衛星。除此之外，當我們的機器變成數量夠多，而且大小齊全的話，我們就可以直接利用其它星球的物質，甚至於從太陽來的物質，實質地創造出成千上萬，就像地球的新星球，沒有人須要被迫去住在地下。

還有除此之外，假使我們選擇要繁衍得夠快和夠久，以致於勢在必行的話，我們還可以跑到其他的星系中。雖然這些可能性看起來似乎怪誕甚或荒唐，但是其實它們只不過是無窮生產能力概念的一些單純的因果而已，而此又是會自我繁衍的智慧型機器概念的一個簡單因果。

因此從長遠看來，我們都不用去掛念成本或是人口壓力。但是現在是我們必須步下奧林帕斯寶座的時刻，而來思考過渡時期中的一些非常真實，而且可能是非常危險的問題。

商業冰凍人的成本

有人可能會認為因著冰凍人的計畫，殯葬的費用就會以相當可觀的倍數激增。然而，現在讓我們來針對這個問題稍加探討一下。

In Detroit, in 1962, according to several leading morticians, funeral costs ranged from about $200 to about $6,000, with an average of perhaps $800. In 1961 the California Funeral Directors Association "suggested" a minimum of $450, and funerals over $1,000 were common. (120)

In Detroit in 1963 a single cemetery plot seemed to cost $80 at least, including perpetual care. (The funds are invested, the interest supplying the maintenance costs.)

In rough figures, then, the total cost of death at present is typically in the neighborhood of $1,000. Now let us try to guess the cost of freezing in the near future, when commercial facilities become available.

The preparation of the body may correspond roughly to a major operation by a team of surgeons using expensive cryogenic equipment, and can therefore perhaps be expected to cost several hundred dollars at least. This might be reduced if mortuary technicians can be trained to replace surgeons.

Even more difficult to assess is the cost of the 'Dormantory' and its maintenance. But there are some suggestive known costs.

In Detroit, in 1963, a mausoleum crypt could reportedly be had for $1,250. The mausoleum itself cost about $3,000,000 to build and holds 6,500 bodies.

在 1962 年的底特律，依據許多個較有名葬儀社的資訊，一般一個葬禮的花費，大約是從 200 美金到大約 6,000 美金不等，其平均花費可能約是 800 美金。在 1961 年，加州葬儀負責人協會 "建議" 一個葬禮最少的費用為 450 美金，而且費用超過 1,000 美金的葬儀，乃是司空見慣的事。(120)

在 1963 年的底特律，一塊單人的墓地似乎至少要花費 80 美金，其中包括了長其的維護。(這筆錢都進入投資，用其所獲得的利息，來提供所需的維護費用。)

因此，以粗略的數字概算，在目前一個葬禮的整個花費，一般地乃是在 1,000 美金左右。現在讓我們試著來概算，在最近的未來，當有商業冰凍設施運轉時，其所需的費用。

身體的準備，大略上可能就相當於由一組外科大夫，用昂貴的冰凍設備，來進行一個大型的手術，因此而可能會被預期至少要花上好幾百塊的美金。假使殮葬的技師可以被訓練來替代外科醫生的話，這個費用可能就可以被減少。

相對上較難估算的乃是 '休眠所，Dormantory' 的費用，以及其維護的費用。但是現在已經有一些已知的建議費用。

在 1963 年的底特律，據說花費美金 1,250 元可以買到一個墓園的墓穴。這個墓園本身的建造成本大約是美金 3,000,000 元，而可以容納 6,500 個遺體。

Can we make a first crude estimate of the cost of a Dormantory by regarding it as a refrigerated mausoleum? Perhaps we can, at least as regards first cost and not maintenance. In fact, since the freezer need not be as fancy nor as spacious as a mausoleum, and need not provide for routine access once it is filled up, possibly its initial cost will be no greater than that of the mausoleum, especially if the refrigeration scheme is the very simple one now to be considered.

To fix a rough upper limit on the cost of maintaining the refrigerating equipment, let us think of the simplest scheme possible; besides being the simplest, it will probably be the cheapest to install and the most expensive to maintain.

This involves merely surrounding the storage space with liquid helium and insulating layers, and replacing the liquid helium as it evaporates.

Now, liquid helium in a 4,000 liter spherical container 2 meters in diameter, shielded by liquid nitrogen, evaporates at about 0.2 percent per day. (103) If we consider a cubical storage space 30 meters on an edge, this will hold 18,000 bodies at 1.5 cubic meters per body. If we assume the evaporation rate is about proportional to the area of the exposed surface, as it ought to be, then the liquid helium evaporating per day would be roughly 3,400 liters.

我們是否可以將休眠所想成是一種冰凍的墓園，而來對其進行一個初步的成本估算呢？至少，在有關初期成本而不管維護成本上，或許我們是可以如此的。事實上，既然，冰凍設施不需要如墓園一樣地華麗和寬廣，而且在其完全進駐後，就不需要提供頻繁的進出，因此其初始的成本將不會比墓園的成本要來得較高，尤其是現在在被思考中的冰凍設計，乃是非常簡約的。

為了要訂定出一個維運冰凍設施的成本上限，就讓我們來看看一個可能是最簡單的設計；這除了是最簡單之外，它在安裝上還可能是最便宜的，但是其維護費用也可能就是最貴的。

這設施包含的只是用液態氦和絕緣層來包裹其儲存空間，以及補充被蒸發消耗的液態氦。

目前，液態氦乃是裝在一個4,000公升的球型容器中，其直徑有2公尺，由液態氮來包覆著，其蒸發速度大約是每天百分之0.2。(103) 假使我們考慮有一個邊長30公尺的一個立方體儲存空間，如果每個軀體佔1.5立方公尺的體積，這空間將可以容納18,000個軀體。假使我們假設，蒸發速度大致上是和暴露表面面積成正比，其時應該也是這樣子的，那麼液態氦每天的蒸發量則大約是3,400公升。

Liquid helium was quoted in Detroit in 1962 at $7 per liter in 100-litre lots. If we use this figure, the evaporation loss cost comes to about $1.32 per day per body, or roughly $480 per body per year. Actually, the price for large amounts will surely be lower. Helium is available in large quantities, occurring as 1 per cent to 8 per cent of natural gas at various wells. (103) On the other hand, we have ignored the cost of replenishing the liquid nitrogen shield; but liquid nitrogen is quoted at only 50 cents per liter in 100 liter lots, and its latent heat of vaporization per dollar's worth is much larger than that of helium, and sufficient insulation could make the heat leak very small, minimizing this cost.

In fact, with very thick insulation, the liquid nitrogen shield could be dispensed with altogether, and the evaporation rate of the helium still reduced, no doubt. In any case, the cost of cooling and recycling the helium will surely be much lower than the cost of simply replacing it, especially after large-scale study and investment. Also, the allotment of 1.5 cubic meters per body may be much too liberal; this is more than 51 cubic feet. All in all, perhaps it is not unreasonable to guess at a figure of $200 per body per year for maintenance as a first approximation.

To produce $200 a year would require capital of $6,667 invested at 3 per cent. (There are always plenty of good bonds for sale which yield this much.) Then adding together the $1,250 storage space cost, the $6,667 capital investment for refrigerating cost, and a few hundred dollars for preparation of the body yields a rough total of $88,500 per body. This is the tentative cost of a private freezer program on a group basis.

在 1962 年，液態氦底特律的報價是以 100 公升的量來出貨，每公升是美金 7 塊錢。假使我們使用這個數字，每個軀體每天蒸發損耗的成本算出來大約是美金 1.32 元，或大約是每人每年美金 480 元。而且事實上如果量大的話，其價格一定會較低。氦氣乃是大量存在的，在各個不同的天然氣井中，都含有百分之 1 到 8 不等的氦氣。(103) 在另外一方面，我們忽略了補足液態氮包覆的成本；然而，液態氮的報價以 100 公升的量來出貨，每公升僅要美金 50 分錢，而且其每塊錢的蒸發潛熱值，比起氦氣來要高出許多，況且只要有足夠的絕緣，就可以讓其熱損耗變得非常地小，因此可以使此成本微小化。

　　事實上，只要用非常厚實的絕緣體，就可以完全免除液態氮包覆的需要，而且無疑地，氦的蒸發速度也會被降低。在任何狀況下，尤其是經過大規模的研究和投資之後，氦氣的循環冰凍成本，將一定會比光是補充氦氣要來得低廉許多。而且，每個軀體 1.5 立方米的配額可能也是過度寬宏的；這是比 51 立方呎還要多。總之，以初步的估算來講，每個軀體每年美金 200 元的猜測數字，或許應該不會是太離譜的。

　　要能夠有每年美金200元的入帳，需要有一筆美金6,667元的本金，在百分之3的報酬率的投資上。(有許多販售中不錯的債券，都有這樣的報酬率。) 將美金1,250元的儲存空間成本，美金6,667元的冰凍成本的投資資本，和幾百塊錢的軀體準備費用加在一起，其所算出的總價大約是每個軀體美金8,500元。此乃是一個以私營冰凍計劃的團體價為架構的一個暫估成本。

It is also interesting to note, for whatever it may be worth, that a 6 cubic foot frozen food locker, holding 150 pounds of meat at a temperature Below 0 F, and of course providing routine access, rents for $10 to $15 per year. (52)

Needless to say, countless refinements and improved safety factors could increase the cost. For example, it might be possible to construct a fully automatic unit with no moving parts, if the Peltier effect can be brought to engineering feasibility. (If an electric current is passed through a circuit containing two different metals, one of the junctions may be cooled and the other warmed; this is the opposite of the thermocouple phenomenon.) Thermoelectric cooling is already receiving considerable attention. (130) The source of power could be thermoelectric as well. Such a sophisticated installation would demand heavy investment, but maintenance might be virtually nil, except for taxes and occasional inspection.

On the other hand, any of many possible developments might reduce the total cost, and tax subsidy might reduce the direct cost.

The Cost of Emergency Storage

If a mutual aid society wanted to store a frozen member in the absence of commercial facilities of any kind, in the immediate future, what might the expenses be?

還有一樣值得注意的事,那就是一個 6 立方呎的食物冰凍櫃,不管其到底有什麼價值,它可以在低於華氏零度下儲存 150 磅的肉品,並且當然有提供時常的進出,而其每年的租金僅是 10 到 15 塊美金。 (52)

在安全因素上許多的精進和改良,將會增高此成本,這當然是不用多說的。例如,假使 Peltier 的效應,其工程可能性可以實現的話,就有可能來建造一個完全自動化的單元,其中不需具有任何會動的零件。(假使將一個電流,流通過一個含有兩種不同金屬的電路,其中一個接點會被冷卻,而另一個則會被加溫;這乃是熱耦現象的反面。) 熱電力冷卻目前已經受到相當的重視。 (130) 電力的來源也可能會是熱電力學的。這類型的精密設施,將會需要大量的投資,然而除了一些定期的和偶發的檢查之外,可能就會完全不需要任何的維修。

在另一方面,在許多可能的研發中之任何一個,都有可能用來降低此總成本,而賦稅上的補助,也可以減少其直接的成本。

緊急儲存的成本

假使有一個互助協會,在沒有商業上的設施運轉之下,要在近期的未來,儲存一個其被冰凍的成員,那這花費可能會多少呢?

Presumably a building would have to be obtained, and caretakers hired, and so on; but what is of interest here is the refrigerating expense.

A rough estimate might be made as follows. Let us assume a container with average dimensions (that is, neither inside nor outside, but in the middle of the insulation) of 7 feet by 3 feet by 3 feet. Let it be metal, with cork board insulation six inches thick. The inside might be divided into a lower compartment, for the body, and an upper compartment, for the refrigerant.

If dry ice is used, other figures entering into the calculation are as follows. The latent heat of vaporization is 246 BTU per pound, the temperature of the dry ice is -109 F, the conductivity of cork board is 0.22 BTU in. per hour per square foot per degree Fahrenheit. (52) The cost of the dry ice is probably less than 6 cents a pound.

If room temperature is taken as 70F, then combining these figures in a simplified calculation gives a refrigeration cost of roughly $4 a day for replacement of dry ice. But this figure can be bettered in many ways.

Even if the crudest methods are used, as sketched above, there will be certain factors working in our favor which are hard to calculate theoretically. For example, the average room temperature will be well below 70F, because it will be unheated in winter, and will be cooled by the carbon dioxide vapor. This effect will be accentuated if there are a number of bodies, in which case there would also be a greater effective average insulation thickness.

設想一下，其中一定要擁有一棟建築物，並且要僱用一些看守人員，還有其它等等；但是，其中最重要的乃是其在冰凍上的開銷。

我們可以做如下的估算。讓我們假設有一個容器，其平均尺寸是 7 英呎乘 3 英呎乘 3 英呎 (也就是不是內部或是外部的尺寸，而是以絕緣體的中間線來量的)。就算它是金屬的，有六英吋厚的軟木板絕緣。其內部可能會被分割成一個下部隔間，來放軀體，以及一個上部隔間，來放冰凍劑。

假使是使用乾冰，其它數字的計算就如下所述。其潛熱的蒸散是每磅 246 BTU，乾冰的溫度是華氏零下 109 度，軟木板的導熱度是每小時每平方英呎每華氏一度 0.22 BTU。 (52) 乾冰的成本可能是少於每磅美金 6 分錢。

假使室內溫度是華氏 70 度，那麼將這些數字結合起來，做一個簡單化的計算，其所獲得的乾冰更換冰凍成本大約是每天美金 4 元。但是這個數字，都可以從許多方面來改進。

甚至於使用上面所描述最粗糙的方法，還是會有某些因素的作用會對我們有利，而這些是難以用理論來計算出來的。譬如說，室內的平均溫度將會是大大低於華氏70度，因為在冬天時將不會使用暖氣，而且將會被二氧化碳蒸氣所冷卻。假使有很多個軀體的話，這個效應將會被更顯著化，在這種情形下，也將會有更大的有效平均絕緣厚度。

Further, if the storage room is in a basement, the earth below and around may provide additional insulation. Also, the previous calculation ignored the heat absorbing power of the carbon dioxide as it warms up, after sublimating, from - 109 F to whatever temperature it reaches before escaping, although this consideration partly overlaps that of room temperature.

If several feet of additional insulation were used, and especially if there were several bodies, it seems to me the cost could easily be reduced by a factor of ten, making it 40 cents per body per day. (The added insulation might be straw or glass wool, the latter being preferable both from the standpoint of insulating quality and of fire hazard. Glass wool is about as good an insulator as cork board.) If there were a sizable number of bodies, and if a specially designed or modified building were used, and if still more insulation were added, the cost might even be brought down to 10 cents daily per body, without the project becoming too unwieldy. We would expect, of course, that in a very few years more economical commercial installations would be come available.

If liquid nitrogen were used, the replacement cost might be over twenty-five times as great. Handling would also be more difficult; but it is not necessary to use gas-tight or pressure-tight containers, except in transport; in fact, an evaporation vent must be provided, as for dry ice.

況且，假使這個儲存室是在地下室，其底下和週遭的土壤也可能提供進一步的隔熱。還有，在前述的計算中，忽略了二氧化碳在暖化過程中，在從華氏零下 **109** 度昇華之後，到其到達其逃逸前之任何溫度的吸熱能力，雖然這個考量會和室內溫度的考量有部分重疊。

　　假使能夠進一步使用好幾呎的絕緣，而且特別是在有好幾個軀體存在之下，依我看來，其成本可以很容易地減少十倍，使其變成每人每天美金 **40** 分錢。(所增加的絕緣物可以是稻草或是玻璃纖維，當然從絕緣品質和火災風險看來，我們會較偏好後者。玻璃纖維大約是和軟木板一樣好的絕緣體。) 假使有相當可觀數量的軀體，而且假使使用一個經過特別設計或是修改過的建築物，並且甚至加入更多的絕緣體，在沒有使此專案變成難以使用之前，其成本甚至還可以被減低到每人每天美金 **10** 分錢。當然我們將可以預期到，在沒有多少年之內，有更多更為經濟的商業設施將會上市。

　　假使所使用的是液態氮，其補充成本將有可能會超過這個的二十五倍。其處置過程也會比較困難；但是除了在運送中之外，此並不見得一定要用到氣密的或是壓密的容器；事實上，就如同在乾冰一樣，一定要提供一個蒸發通氣孔。

Trust Funds and Security

Before submitting to freezing, people will make strenuous attempts to safeguard their dormant bodies, and to ensure firm positions in society on revival. It may be expected, for example, that elaborate trust funds will be set up.

Those who try to "take it with them" will want reliable supervision of the freezers, and will hope, through the magic of compound interest, to awaken wealthy. Yet at first thought one is apt to doubt that everybody can awaken rich, because this is somehow "against nature," or would represent "something for nothing." We also realize that future governments could confiscate any property and outlaw any trust arrangements at will.

While the considerations involved are very complex, both economically and psychologically, and predictions can be only half educated guesses, still there are some pertinent remarks to be made.

Interest rates depend, of course, on two broad factors, one physical and the other psychological. The first concerns the productivity of a dollar, that is, the rate of production of wealth by a dollar's worth of capital goods. The second relates the supply and demand situation. The physical productivity factor, of course, one expects to increase continuously, but the psychological factor almost defies analysis, let alone prediction.

信託基金和證券

在被交付冰凍之前，人們會積極地努力，試著要來保全他們在休眠中的身體，而且要來保障他們在復甦之後社會地位的穩固。可預期地，一定會有例如像是設計縝密的信託基金的被設立。

那些試著要"隨身攜帶著"的人，將會要求對冰凍器可靠的監控，而且將會希望通過神奇的複利，使其醒來時是富有的。但是在第一個念頭閃過時，一個人很容易就會對每一個人都可以醒來富有產生懷疑，因為這是有點"違背自然"，或是代表著"無中生有"。我們也體察到，未來的政府有可能會肆意地來充公任何的財產，並且會非法化任何的信託安排。

雖然這些顧慮所牽扯的，不管在經濟學上或是在心理學上，都是非常地複雜，而且一些預測可能都會淪於半調子的學術猜測，但是還是有一些相關的論點應該要被提出。

利息當然會和兩個大因素有關，一個是物質的，另外一個則是心理的。第一個是和一塊錢的生產力相關，那就是一塊錢價值的資本財產生財富的速率。第二個則是和供給和需求的狀況相關。當然一個人可以期待物質生產力因素會持續地增加，但是心理上的因素則幾乎是無法分析的，更不用想去預測。

If this is correct, one can do little except to take experience as a rough guide, without trying to estimate the effect on the money market of the supply represented by the trust funds.

Let us ignore taxes, and assume we can hedge against inflation by always having part of the money in equity investments. Then, if the return on conservative investments is something like 3 per cent yearly during the freezer era, $1,000 untouched would grow to roughly $19,000 in 100 years, $370,000 in 200 years, or $7,000,000 in 300 years. This money is real; it represents initially the diversion of buying power from consumer goods to capital goods, followed by continuous reinvestment. If such wealth seems awesomely huge, we must remember that the productivity par capita of the nation is now increasing by almost 2.5 per cent yearly, and the rate of increase will probably improve greatly. The annual Gross Product per capita in the year 2264 of what is now the United States, even if the rate of growth does not improve, will be about $4,500,000! In 1960 it was only about $2,800. (49)

I see no reason to expect future generations to be jealous of the bank accounts and financial influence of the frozen. Those breathing will get later starts in saving, but will be able to save from much larger incomes, and would not have to be second class citizens financially, even without discriminatory or confiscatory laws aimed at the frozen.

假使這是正確的，那麼一個人除了以經驗為粗略的指引之外，就不能做什麼了，他沒有辦法預測由信託基金所代表的供給面，來試著估算出其對貨幣市場的影響。

　　讓我們先將稅賦忽略，並且假設藉著永遠有一部分的金錢，是放在產權的投資上。那麼假使這樣保守的投資報酬率，在其冰凍的期間每年大約有百分之 3，在 100 年內，沒有被動到的 1,000 元大約將會成長至 19,000 元，在 200 年內，會成長至 370,000 元，或是在 300 年內，會成長至 7,000,000 元。這些錢是真實的；其起初所代表的，乃是將購買力從消費財轉移到資本財，然後接著又是連續地再投資。假使這種財富看起來似乎太過龐大，我們必須要切記，現在這個國家的每人生產力，幾乎是以每年百分之 2.5 在增加，而且這個增加的速率，將有可能會大大地被再提昇。現在美國每個國民的每年生產毛額，就算其增加速率沒有提昇，到 2264 年大約將會達到 4,500,000 美元！在 1960 年時，大約僅有區區 2,800 美元。(49)

　　我實在看不出有什麼理由會使得未來世代的人，來忌妒冰凍者的銀行存款戶頭和其在財務上的影響力。就算在沒有針對冰凍者的歧視性或是充公性的法律之下，這些活著的人只是在儲蓄上將會晚一點起步，但是卻將可以從多很多的收入中來儲蓄，並且不一定要淪為財務上的二等公民。

The people in the freezers should also be protected by family loyalties, and by a tradition which recognizes that each in his turn (until the generation that achieves immortality) must become frozen and helpless, dependent on the good will and law-abiding character of his successors.

It must also be remembered that before long the option of suspended animation will be available. Some individuals will choose cold sleep before they become senile, and will therefore be able to arrange for periodic awakenings to look the situation over and check up on Junior.

It is not easy to anticipate the legal and sociological consequences of these visits by great-grandpa. Some of us might feel a little queasy at the notion, so to speak, of a zombie climbing the cellar steps every few years, with the frost in his beard, to cast a fishy eye on the family and perhaps vote his shares at the election of directors of an important corporation. But one grows accustomed to everything, and it rather seems the net result could be a beneficial tradition of permanence of the family and institutions, a strengthened feeling of the unity of mankind, an ingrained sense of our endless responsibility for each other.

World Relations

Very little has been said so far about the prospect of immortality as seen from elsewhere than the United States, Europe, and similar regions. How will it affect the internal and external policies of the retarded nations? Of the communists?

這些在冰凍器中的人，也一定要能夠受到家族忠誠度和傳統的保護，而且此乃是基於每一個人有一天，一定都會輪到他變成冰凍而且無助的時候的認知 (直到能夠達到永生不死的世代為止)，這乃是有賴於他後代的誠意和遵守法律的人格。

也務必要切記的，再不久將會有活體休眠的選擇可能性。有些個人將會在其變成風燭殘年之前，選擇冰凍休眠，如此一來，他將可以安排週期性的甦醒，來觀看情勢並且看看他的後代。

要能夠預期這些曾曾祖父的拜訪，所可能產生出來的法律上和社會學上的後果，並非是輕而易舉的事情。一想到這個，就好像說，一個僵屍每隔幾年就會攀爬地窖的階梯，在其鬍鬚上還帶有冰霜，對其家族成員投射出呆滯的眼神，而且可能會在一家重要公司的董事選舉中，投下他的股權，我們當中某些人可能就會因此而感到有一些反胃。但是，人總是會漸漸地適應各種事情的，而且這最後的結果看起來，反而可能會是有助益於家族和機構傳統的永續，一個全人類一體感覺的增強，一個我們對每一個人之間無止境的責任感的加深。

世界關係

到目前為止，我們很少提及在美國，歐洲和類似區域之外，從其它地方來的，對有關永生不死期盼的觀感。它將會如何影響一些發展較為遲滯國家，其對內和對外的政策呢？會如何影響共產主義者呢？

At this juncture perhaps little can be said of what will actually develop, if something can be said about what could happen and ought happen.

The first reaction on the part of leaders of backward or totalitarian states might be unfavorable, since a freezer program could put heavy extra demands on already inadequate resources and could weaken discipline by substituting materialistic goals for the quasi-religious ideals of the self-styled revolutionaries. To help clarify the problem, a few remarks about the nature the "emerging" nations and the "communist" nations may be in order, representing common knowledge which is not always made explicit.

Economically, they most often stand for a kind of socialist state capitalism, and in this respect differ little from some Western countries. Politically, they usually represent bureaucracy enthroned, or, to use an older word meaning almost the same thing, oligarchy, and in this respect also represent nothing new in the world, and differ little from many established and rightist countries. Even their usual totalitarian character, the enshrining of the state high above the individual, shows no radical departure either from earlier history or from various established states of the political right. The driving force, in the case of the undeveloped countries, is largely just racist or nationalist patriotism or chauvinism, distilled into a foggy ideal. In the case of the Reds there is an additional unifying mystic element based on the Word of the Prophets, Marx and Lenin. From the standpoint of the leaders, the goal may be personal and national aggrandizement, and the "ideologies" may be only tools to pry obedience and self-sacrifice out of the people.

在這個時空交會點，真正到底將會怎麼發展，我們或許沒有辦法說什麼，假使可以說些什麼的話，應該是有關可能會發生什麼，以及什麼應該要發生。

　　部分落後或是極權國家的領導人，其第一個反應可能會是反對的，因為一個冰凍計劃，可能會在其資源已經相當不足的困境上，加諸極重的需求，而且可能會弱化其以偽宗教理念自成一格的革命精神，而來替代之以物質主義為目標的訓誨。為了要搞清楚這個問題，可能必須要提出一些有關"新興"國家和"共產主義"國家的詮釋，來替代一些常常是沒有被解釋清楚的普通常識。

　　在經濟學上，他們最常代表的乃是一種社會主義國家的資本主義，而在這一方面，是和一些西方國家沒有多大的不同。在政治學上，他們通常是代表著官僚掛帥，或是用一個意義幾乎相同的老字眼來說，就是寡頭統治，而在此方面其所代表的，在世界中也並非是新鮮事，而和許多已存在的以及右派主義國家沒什麼兩樣。甚至於他們通常的獨裁性格，將國家高高供奉在個人之上，這也沒有顯示出極端地從先前的歷史，或是從各種不同政治右翼的現有政體中偏離。以未開發的國家案例來講，其驅動力大部分僅是種族主義者，或是國家主義者的愛國心或是沙文思想，被蒸餾成一種模糊的理想。在紅色共產的案例中，其中還多了一種架構於馬克思和列寧，先知的話語上的一種統一的神秘元素。從這些領導者的觀點看來，其目標可能是個人的和國家的擴張，並且這些"意識型態"可能僅是用來從人民中壓榨出順服和自我犧牲的工具而已。

Words have an amusing way of becoming twisted in usage, and while it is commonplace to regard ourselves as idealists and the Reds as materialists, in fact the reverse is nearer the truth. We are mature enough to be materialistic in the sense of wanting freedom and wealth for ourselves, and not just for some dim posterity, and in the sense that we try to remember the state is only an instrument of the people, only a means to an end. The Reds, on the other hand, are childish idealists to the extent that they are willing to sacrifice themselves for slogans and embrace a kind of mysticism in imbuing their state and ideology with intrinsic worth and permanent meaning. It is we who are generally godless and not the Reds: we may acknowledge the ascendancy of Jehovah, but seldom consult Him in practical affairs, whereas they pay a more sincere homage to their god-in-overalls, through his prophets, Marx and Lenin, looking to them for day-to-day guidance. Soviet workers are so pious that they have sacrificed their right to strike on the altar of Marxism-Leninism.

Serious dangers therefore arise. Many leaders of the eastern and southern countries may feel a freezer program would threaten the very foundations of their regimes. The people themselves, who often take pride in the term "revolutionary" but in fact may notably lack intellectual flexibility and adaptability, may find it difficult to switch gears and reorient themselves. A fury of bafflement, resentment, and jealousy may even exacerbate international tensions at first. But there are hopeful factors as well.

在使用文字上有一種很好玩的方法，來使其變成意義上的扭曲，而當認為我們自己才是理想主義者，而紅色共產則是唯物主義者，變成非常普遍時，事實上其反面才較接近真理。在為我們自己的自由和財富方面的爭取，而不僅是為某些語焉不詳的後代子孫爭取，以及在我們努力去記住，國家僅是一種人民的工具，僅是達到目的的一個手段方面，我們已經足夠成熟來變成是唯物辨證的。在另一方面，紅色共產人士都是童騃的理想主義者，他們竟然到一個地步，願意為口號而來犧牲他們自己，並且竟然會去擁抱在用內在價值和永恆意義，將他們的國家和意識型態浸透後，所產生的一種神秘主義。我們才是一般的無神主義者，而不是紅色共產人士：我們可能會認知到耶和華的優越，但是在實際事物上，卻是很少去諮詢祂，然而他們卻對他們統管一切的神，獻出更加虔誠的尊禮，透過他們的先知們，馬克思和列寧，仰賴他們來獲得每日的導引。蘇俄的工人是虔誠到可以犧牲他們的權利，來到馬克思和列寧的祭壇上膜拜。

因此，會產生出一些嚴重的危險。東方和南方國家中的許多領袖，可能會感覺冰凍計劃將會直接威脅到他們政權的根本。這些人，他們自己常常以"革命性"的字眼來引以為傲，但是事實上，有許多人都明顯地缺乏智慧的伸縮度和適應力，有許多人都發覺自己無法轉換車檔，重新調整自己的方向。一種由困惑，怨恨和妒忌所釀成的憤怒，在一剛開始，甚至於可能會使國際間的緊張惡化。然而其中畢竟還是會存在有一些希望的因子。

The communists, and even their leaders, are after all not demons, but people like ourselves, struggling to live in, and make sense out of, a very difficult and mysterious universe. Desperation makes fanatics, but hope - on a practical, personal level - may be the key to cooperation.

The nationalist and leftist leaders may buzz angrily about for a while, like hornets in a bottle, but they should quiet down as they come gradually to realize two things. First, they will want immortality for themselves and their families. Second, all problems take on a completely different perspective in the long view. When the future expands, the past shrinks; historical affronts lose their sting, and vendettas their fascination. The words of the song then make self-evident good sense, that is, to eliminate the negative and accentuate the positive.

Many compromises and makeshifts may be necessary to stretch the rupees, pesos, etc. For a time the strictest economies in freezing may have to be practiced in many countries. Perhaps bodies will be stored in pits insulated with straw and cooled with dry ice. It is even possible that after freezing with dry ice they will be shipped to Siberia for natural cold storage, if it is decided that the changes at these temperatures are limited, or that the cost of maintaining artificially low temperatures here is sufficiently less to warrant the cost of transportation. From the standpoint of civil order, it will not at first greatly matter how skillfully the bodies are preserved, so long as hope is preserved.

共產主義者，甚至於他們的領袖，畢竟都不是魔鬼，而是和我們一樣，他們都還是人，在一個非常困難和神秘的宇宙中，試著要掙扎來求生存，並且要去理出其頭緒。絕望是會產生一些狂熱者，但是希望－在一個務實的，個人的層次上－可能會是促成合作的關鍵。

這些國家主義者和左派的領袖們，有一陣子可能會到處生氣地亂叫，就好像被關在瓶子中的黃蜂，但是當他們慢慢地去理解到兩件事情之後，他們就應該會安定下來。第一件，他們將會要為他們自己和他們的家人爭取永生。第二件，以長遠的觀點看來，所有的問題都會變成一種完全不同的觀感。當未來擴張開來時，過往的就會萎縮掉；歷史上的一些衝突冒犯，都將會喪失它們的刺痛力，一些種族仇殺，都將會喪失其魔幻力。於是詩歌中的字句，就變成自我證實的好話，那就是，要去消滅掉負面消極的，並且要去提倡正面積極的。

可能會需要許多的妥協和權宜之計，才有辦法使盧比，披索，等等增值。會有一段時間，在許多國家中，可能一定會發生在冰凍上所面臨最嚴重的經濟狀況。軀體將有可能要被儲存在用稻草隔熱和用乾冰冰凍的一些土坑之中。在用乾冰冰凍過後，假使西伯利亞溫度的變化，被認定為是有限的，或是在此以人工方法維護低溫的成本，低到足夠省下運輸的成本的話，他們甚至有可能要被運送到那裡，去獲得天然的冰凍儲存。以市民優先秩序的觀點看來，他首先將不會介意這些軀體被保存的技術有多好，只會希望有被保留下來就好。

Demands will increase with time and learning, but so, one hopes, will resources and cooperation. In particular, this jolt may abruptly shift the birth control program into high gear. One may even dare hope that before too long the poorer countries will prefer cryobiological aid to military aid. There are perils in plenty, but there is also much room for optimism.

隨著時間和知識的增加，其要求將會增加，既然如此，我們也期望資源和合作也將會因此而增加。特別是，這個震撼可能會突然間地，將生育控制計劃轉進入高檔。我們可能甚至於敢去預期，在不久的未來，較為貧窮的國家，將可能會較偏愛冰凍生物學上的援助，而勝過於軍事上的援助。雖然其中的危機重重，但是畢竟也還有許多空間來樂觀的。

CHAPTER VIII

The Problem of Identity

身分上的問題

CHAPTER VIII

The Problem of Identity

In considering the chances of reviving, curing, rejuvenating, and improving a frozen man, we have to envisage the possibility of some very extensive repairs and alteration. This leads to a number of very perplexing puzzles.

As an extreme case, imagine an elderly cancer victim who is not frozen until several hours after death, and then only by crude methods. Almost all the cells of his body have suffered severe damage and are thoroughly dead by present criteria, although some would grow in culture and we assume a small percentage of them have degenerated relatively little. But after enough centuries pass medical art at last is ready to deal with him, and for the sake of emphasis let us assume a grotesque mixture of techniques is used.

When our resuscitee emerges from the hospital he may be a crazy quilt of patchwork. His internal organs - heart, lungs, liver, kidneys, stomach, and all the rest - may be grafts, implanted after being grown in the laboratory from someone else's donor cells. His arms and legs may be bloodless artifacts of fabric, metal and plastic, directed by his own will and complete with sense of touch but extended and flexed by tiny motors. His brain cells may be mostly new, regenerated from the few which could be saved, and some of his memories and personality traits may have had to be imprinted on or into the new cells by microtechniques of chemistry and physics, after being ascertained from the written records.

第八章

身分上的問題

在思考復活、治癒、回春和改良一個冰凍人的各種機會時，我們一定要能夠預先看到一些非常大範圍的修復和變更的可能性。因為這些會導引出一些非常令人困惑的謎題。

舉一個極端的例子來看，想像有一個老年的癌症患者，他是在死後好幾個小時,才被冰凍起來,而且僅是用非常粗糙的方法冰凍的。他身體中全部的細胞幾乎都蒙受了嚴重的傷害，而且以目前的標準來看，是已經完完全全死了，雖然其中某些細胞還是會在培養皿中成長，而且我們也認為它們之中會有一小部分僅受到相對微小的破壞。但是經過好幾百年足夠長的時間之後，醫學的技術終於已經成熟到可以來處理他了，而且為了要強調這個，讓我們假設會有一大堆新奇技術的混合應用。

當被我們復甦的人從醫院出現時，他可能會像是一塊怪異補丁的毯子一樣。他體內的一些器官、心臟、肺臟、肝臟、腎臟、胃和所有其它的，都有可能從某一個人所捐贈的細胞，在實驗室中培養之後而接植，移植而來的。他的手臂和腿腳可能是布纖、金屬和塑膠作成的無生命的人造物，而且配備有觸覺功能，可以藉由他自己的意志來控制，其伸展和收縮則由微小馬達來執行。他的腦部細胞大部分可能都是新的，它們乃是從那少數幾個可以被保存的腦細胞再生出來的，而且他的某些記憶和個人的特質，可能必須是在經過文書紀錄的確認之後，藉著化學和物理的微技術，來將之灌入或是印入這些新的細胞。

Striding eagerly into the new world, he feels like a new man. Is he?

Who is this resuscitee? For that matter, who am I and who are you?

Although most resuscitees will not represent such extreme cases - we hope most of us will be frozen by non-damaging methods - nevertheless we cannot sidestep the issue. We are now face to face with one of the principal unsolved problems of philosophy and/or biology, which now becomes one of prime importance in an exceedingly practical way, namely that concerning the nature of "self."

What characterizes an individual? What is the soul, or essence, or ego? This seemingly abstruse question will shortly be seen to have ramifications in almost every area of practical affairs; it will be the subject of countless newspaper editorials and Congressional investigations, and will reach the Supreme Court of the United States.

We can bring the problem into better focus by putting it in the form of two questions. First, how can we distinguish one man from another? Second, how can we distinguish life from death?

Later I shall offer some tentative partial answers. First we can illuminate the question, and perceive some of its difficulties and subtleties, by considering a series of experiments. Some of these experiments are imaginary, but perhaps not impossible in principle, while others have actually been performed.

他感覺到自己像是一個新人，迫不及待地想大步踏入這個新的世界。但是，是他嗎？

這個被復甦者到底是誰？為了這個，我們也要問，我到底是誰，你到底是誰？

雖然這些極端的案例將不會代表著大部分的被復甦者－我們也希望我們大多數都可以用非破壞性的方法來被冰凍－然而，我們卻不能將此問題擱置不顧。我們現在正和一個哲學上以及／或是生物學上的一個主要未解的問題面對面，而此在非常的現實面上，已經變成一個超級重要的問題了，也就是它牽扯到了"自我"的本質。

什麼是一個人的特質呢？什麼是靈魂，或是存在，或是自我呢？這個看起來似乎深奧難解的問題，我們不久將會看到它幾乎和所有現實事物中的每一個領域都有關聯；它將會是無數報章評論和國會調查的主題，而且層級將會高達美國的最高級法院。

我們可以將此問題以兩種問題的型式來表達，以便更能較清楚地抓到問題的焦點。第一，我們如何來分辨出一人與另一人呢？第二，我們如何來分辨出生命和死亡呢？

稍後我將會提出某些暫時性的部分解答。首先，我們可以藉著思考一序列的實驗，來看清楚這個問題，並且來察覺存在其中的困難點和微妙處。雖然這些實驗中，有些是想像的，但是在原理上，可能並非是不可能的，而其它的，在實際上則已經被進行過了。

Experiment 1. We allow a man to grow older

Legally, he retains his identity; and also subjectively, and also in the minds of his acquaintances (usually). Yet most of the material of his body is replaced and changed; his memories change, and some are lost; his outlook and personality change.

It is even possible that an old acquaintance, seeing him again after many years, might refuse to believe he is the same person. On first considering this experiment, we are apt to feel slightly disturbed, but to retain a vague conviction that "basically" the man is unchanged. We may feel that the physical and psychological continuity has some bearing on the question.

Experiment 2. We watch a sudden, drastic change in a man's personality and physique, brought about by physical damage, or disease, or emotional shock, or some combination of these. Such has often occurred.

Afterwards, there may be little resemblance to the previous man, mentally or physically. There may be "total" amnesia, although he may recover capability of speech.

Of course he retains, e.g., the same fingerprints, and the same genes. But it would be absurd to say the main part of a man is his skin; and identical twins have the same genes, yet are separate individuals.

Although the physical material of his body is the same stuff, he seems – and feels - like a different person. Now we are more seriously disturbed, because the main continuity is merely physical; there is a fairly sharp discontinuity in personality.

實驗 1。我們容許一個人可以變老

在法理上，他會保留他的身分；而且在主觀上，以及 (通常地) 在他認識的人的心靈當中，也是如此。但是他的身體中大部分的物質，都已經被替代過和改變過了；他的記憶會改變，而且有些會喪失；他的外表和個性都會改變。

可能會有一個老朋友，經過許多年之後，再度看到他時，甚至於可能會拒絕去相信他就是所認識的同一個人。剛一想到這個實驗時，我們很容易地都會感到些微的困擾，但是還是會保留住一份模糊的信念，認為這個人"基本上"是沒變的。我們可能會覺得，其身體上和生理上的連續性是和這個問題有相關的。

實驗 2。我們看到一個人的個性和體型會因為身體的傷害或是疾病，或是情緒上的打擊，或是其中某些組合，而造成突然和強烈的變化。這種事情已經常常在發生了。

在事發之後，不管是在心理上或是在身體上，可能是和其先前之人已經大相逕庭了。雖然他可能恢復了說話的能力，但是他可能已經"完全地"喪失記憶了。

當然他還是會保留有，例如像是同樣的指紋，以及同樣的基因。但是如果說一個人的主要部分乃是在其表皮，那倒是有點荒謬；而且一對同卵雙胞胎也具有相同的基因，但是卻是兩個完全不同的個體。

雖然他身體的實際構成物質還是那堆同樣的東西，但是他看起來似乎—而且感覺起來—像是一個完全不同的人。現在我們則是更加嚴重地感到困擾了，因為其中所存在的連續性僅僅是身體上的；在人格上，卻存在著一個相當明顯的不連續點。

One might say with some plausibility that a man was destroyed, and another man was created, inheriting the tissues of his predecessor's body.

Experiment 3. We observe an extreme case of "split personality."

It is commonly believed that sometimes two (or even more) disparate personalities seem to occupy the same body, sometimes one exercising control and sometimes the other. Partly separate sets of memories may be involved. The two "persons" in the same body may dislike each other; they may be able to communicate only by writing notes when dominant, for the other to read when his turn comes.

We may be inclined to dismiss this phenomenon by talking about psychosis or pathology. This tendency is reinforced by the fact that apparently one of the personalities is usually eventually submerged, or the two are integrated, leaving us with the impression that "really" there was only one person all along. Nevertheless, the personalities may for a time seem completely distinct by behavioral tests, and subjectively the difference is obviously real. This may leave us with a disturbing impression that possibly the essence of individuality lies after all in the personality, in the pattern of the brain's activity, and in its memory.

Experiment 4. Applying biochemical or microsurgical techniques to a newly fertilized human ovum, we force it to divide and separate, thereby producing identical twins where the undisturbed cell would have developed as a single individual. (Similar experiments have been performed, with animals.)

我們甚至可以以具有相當程度的可信度來說，這個人乃是死了，而另外一個人被創造了，來承繼先前那個人身體中的各種組織。

實驗 3。我們觀察一個"人格分裂"的極端案例。

大家都普遍地相信，有些時候兩個（或是更多個）相異的人格，似乎可以佔據同一個身體，有些時候其中一個會施展控制，有些時候則由另外一個來控制。這可能是牽扯到了部分分離的記憶套組。這兩個在同一個身體中的"人"還可能相看兩相厭；他們之間可能僅能靠著其中一個在主導時所留下的字條，讓輪到另外一個主導時，可以讀取而來溝通。

我們或許會藉著精神病學或是病理學的討論，而傾向於對此現象嗤之以鼻。很明顯地，因著其中的一個人格通常都會消失，或是這兩個有時會整合成一的事實，而讓這種傾向得以加強，留給我們的印象就是，一路走來"事實上"就只有一個人而已。但是不管如何，由行為測試看出，這些人格曾經看起來是完全地不同，而且主觀地看來，很明顯地這個差異是真實的。這個可能會留給我們一個煩擾的印象，畢竟，構成個人的實質要素可能就是在於其人格，在於其腦部活動的模式，和在於其記憶。

實驗4。利用生物化學或是顯微手術技術，在一個剛剛授精的人類卵子上，我們可以強迫它來分裂而分離，因而可以產生出同卵雙胞胎，而沒有經過騷擾的細胞則只會成長出一個單一的個體。（同樣的實驗已經在動物上面試驗過了。）

An ordinary individual should probably be said to originate at the moment of conception. At any rate, there does not seem to be any other suitable time - certainly not the time of birth, because a Caesarean operation would have produced a living individual as well; and choice of any other stage of development of the fetus would be quite arbitrary.

Our brief, coarse, physical interference has resulted in two lives, two individuals, where before there was one. In a sense, we have created one life. Or perhaps we have destroyed one life, and created two, since neither individual is quite the same as the original one would have been.

Although it does not by any means constitute proof, the fact that a mere, crude, mechanical or chemical manipulation can "create a soul" suggests that such portentous terms as "soul" and "individuality" may represent nothing more than clumsy attempts to abstract from, or even inject into, a system certain "qualities" which have only a limited relation to physical reality.

Experiment 5. By super-surgical techniques (which may not be far in the future) we lift the brains from the skulls of two men, and interchange them.

This experiment might seem trivial to some. Most of us, after thinking it over, will agree it is the brain which is important, and not the arms, nor the legs, nor even the face. If Joe puts on a mask resembling Jim, he is still Joe; and even if the "mask" is of living flesh and extends to the whole body, our conclusion will probably be the same. The assemblage of Joe's brain in Jim's body will probably be identified as Joe. But at least two factors make this experiment non-trivial.

一個普通正常人的生命，可能應該被說成是在授精的那一刻開始的。不管如何，似乎再也沒有其他較恰當的時刻了－當然不會是指在誕生的那一刻，因為剖腹生產手術也是可以生產出一個活生生的個體；而如果選在其它胚胎成長的任何一個階段，則又有點隨意亂選的。

我們簡單的，粗糙的實體干擾，會導致兩條生命的產生，也就是兩個個體的產生，而不然的話就僅會有一個。換句話說，我們已經製造出一條生命。或許也可以說我們破壞了一條生命，而製造出兩條生命，因為這兩條生命中沒有一條是會和本來的那一條一樣的。

雖然不管如何，這是沒有任何證據可以來證明的，但是藉著一種僅是粗糙的，機械的或是化學的操作就可以"創造出一條靈魂"的事實指出，這些如"靈魂"和"個性"的一些神奇的詞彙，所代表的可能僅是想要從一個系統中榨取，甚或是灌入某些"品質"的一些齷齪的嘗試而已，而這些虛幻的質感和實際的現實之間卻僅有非常少的關聯。

實驗 5。利用超級手術技術（這可能在不遠的未來就有可能），我們將兩個人的腦部從其頭顱中取出，然後將其互相對調。

這個試驗對某些人而言，可能似乎是沒有什麼。在經過仔細思考過後，我們大部分的人將會同意，大腦才是最重要的，而不是雙手，也不是雙腳，甚至於也不是臉。假使阿喬戴上一個像是阿金的面具，他還是阿喬；而且甚至於這個"面具"是由活的血肉作成的，並且覆蓋了整個身體，我們所下的結論可能還會是一樣的。以阿喬的頭腦和阿金的身體所組成的組合體將可能會被認定為是阿喬。但是其中至少有兩個因素讓此試驗非比尋常。

First, if the experiment were actually performed and not merely discussed, the emotional impact on the parties concerned would be powerful. The wives would be severely shaken, as would the subjects. Furthermore, Joe-in-Jim's-body would rapidly change, since personality depends heavily on environment, and the body is an important part of the brain's environment. Also, we may be willing to admit that Joe's arms, legs, face, and intestines are not essential attributes of Joe - but what about his testicles? If Joe-in-Jim's-body lies with one of their wives, he can only beget Jim's child, since he is using Jim's gonads. The psychiatric and legal problems involved here are formidable indeed.

Some people might be tempted to give up on Joe and Jim altogether, and start afresh with Harry and Henry. In one sense, this is an impractical evasion, since the memories, family rights and property rights cannot be dismissed. From another view, it may be a sensible admission that characterization of an individual is to some extent arbitrary.

Once again, the suggestion is that physical systems (i.e., real systems) must in the end be described by physical parameters (operationally) and that attempts to pin profound or abstract labels on them, or to categorize them in subjective terms, cannot be completely successful.

Experiment 6. By super-surgical techniques (not yet available) we divide a man's brain in two, separating the left and right halves, and transplant one half into another skull (whose owner has been evicted).

Similar, but less drastic, experiments have been performed.

第一，假使這個實驗真的進行了，而不光光是被討論而已，那麼對所牽扯入的各方的人，其感性上的衝擊將會是非常地巨大。就和其主角一樣，他們的老婆也將會受到極大的震撼。況且，在阿金身體中的阿喬將會改變得非常快，因為個性和環境的關係非常密切，而身體乃是大腦環境中的一個重要的部分。而且，我們或許會願意承認，阿喬的雙手，雙腳，顏面和腸道都不是阿喬的重要部位，但是他的睪丸是不是呢？假使阿金身體中的阿喬和他們其中的一個老婆上床時，他僅能生出阿金的小孩，因為畢竟他是使用著阿金的生殖腺。這裡所牽扯的精神病學上和法學上的問題的確是相當可怕的。

有人可能會想乾脆去將阿喬和阿金完全放棄，而以阿哈和阿亨重新來過。但是從某一觀點看來，這僅是一種不務實的逃避，因為畢竟所有的記憶，家族權益和財產權利都是無法否認的。而從另一個觀點看來，將一個個人的特徵，在某一程度上，認為是可以任意亂定的乃是合理的。

再度強調，我們的建議就是，一些實質的系統（也就是真實的系統）到最後一定要以實質的參數（可以運作地）來描述，而那些要將深遠的或是抽象的標籤，標在它們上面的嘗試，或是要用一些主觀的詞彙，來將它們歸類的，都將會是完全地徒勞無功的。

實驗 6。利用超級手術技術（這是目前還沒有的），我們將一個人的頭腦分成兩半，再將其分離成左半部和右半部，並且將其中各部的一半移植到另一邊的頭顱（他的主人早就被驅逐出門了。）

類似這樣，但是較不那麼極端的一些實驗，已經有被進行過了。

- 375 -

Working with split-brain monkeys, Dr. C. B. Trevarthen has reported that " . . . the surgically separated brain halves may learn side by side at the normal rate, as if they were quite independent." (121) This is most intriguing, even though the brains were not split all the way down to the brain stem, and even though monkeys are not men.

There is also other evidence in the literature which we can summarize, with certain simplifications and exaggerations, as follows. Either half of a brain can take over an individual's functions independently. Normally, one half dominates, and loss of the other half is not too serious. But even if the dominant half is removed, or killed, the other half will take over, learning the needed skills.

There is presently no conclusive evidence that so drastic an experiment as ours would necessarily succeed; but in principle, as far as I know, it might, and we are not at the moment concerned with technical difficulties.

If it did succeed, we would have created a new individual. If the left half was dominant, we might label the original individual LR; the same skull containing the left half alone after surgery we might call L, and the right half alone, in a different skull after the operation, is R. L thinks of himself as being the same as LR. R may also think of himself as LR, recuperated after a sickness, but to the outside world he may seem to be a new and different, although similar, person.

In any case, R is now an individual in his own right, and regards his life to be as precious as anyone else's. He will cling to life with the usual tenacity, and if he sees death approaching will probably not be consoled by the knowledge that L lives on.

以一些腦部被分離的猴子做實驗，C. B. Trevarthen 博士發表說，"...這些被手術分離的腦的兩半部，都會以一般正常的速度齊頭並進地來學習，它們就好像是相當獨立的一樣。"(121) 雖然這兩部分的腦是沒有全部地被分離到其腦幹，而且雖然這是猴子而不是人，但是這還是非常地引人入勝的。

在科學文獻中還有一些其他的證據，我們可以用稍微簡化和誇張的方法來將之簡縮如下。一個頭腦的任何一半，都可以獨立地來接掌一個人的各種功能。在一般狀況下，都是由一半來主導的，因此失去另外一半都不會是太嚴重的問題。但是就算是主導的那一半被移除了，或是殺死了，其它另外一半就會接手，而且會去學習所需要的各種技能。

目前是沒有任何具結論性的證據，可以用來證明像我們那麼極端的實驗一定是會成功的；但是據我所知，在原理上這是有可能的，我們在此時此刻並不想要去考慮其技術上的各種障礙。

假使這真的成功了，那麼我們就會創造出一個新的個體。假使先前是左半部在主導，我們可以將其本來的個體標誌為 LR；而在手術後，對同一個頭顱而僅包含有左半部的，我們可以稱之為 L，對手術後放在一個不同頭顱中僅有右半部的，我們可以稱之為 R。L 會認為他自己就是和 LR 一樣。R 也可能會認為他自己就是從生一場病後恢復過來的 LR。他雖然看起來似乎是有點相像，但是就外在世界的人而言，卻是一個全新而且不同的人。

不管如何，R現在是一個有他自己權利的個體，而且認為他的生命是和其他人的一樣地寶貴。他將用同樣堅強的力量來擁抱生命，而且假使他看到自己死亡將至時，將可能也不會因為知道L還活著，而得到慰藉。

Even more interesting is the attitude of L, the formerly dominant half, now alone in the skull. Suppose that, before the operation, we had told LR that the dominant half of his brain was diseased, and would have to be removed, but that the other half would take over, albeit with some personality changes and possibly some loss of memory. He would be worried and disturbed, certainly -- but he would probably not regard this as a death sentence. In other words, LR would be consoled well enough by the assurance that R would live on. Yet after the splitting, and transplanting operation, L would regard his own destruction as death, and it would not satisfy him that R lived on, in another body.

This experiment seems to suggest again that, psychologically if not logically, the physical continuity is an important consideration.

Experiment 7. A man is resuscitated after a short period of clinical death, with some loss of memory and some change in personality.

This experiment has actually been performed many times. (97) Death was real by the usual clinical tests (no respiration, no heartbeat) but of course most of the cells remained alive, and most people would say that he had not "really" died, and that he was certainly the same person afterward. This experiment is important only as background for the following ones.

Experiment 8. A man dies, and lies unattended for a couple of days, passing through biological death and cellular death. But now a marvel occurs; a space ship arrives from a planet of the star Arcturus, carrying a supersurgeon of an elder race, who applies his arts and cures the man of death and decay, as well as his lesser ailments.

更有趣的就是有關 L 的態度，這就是先前在主控的半部，現在則是單獨地存在在頭顱當中。假使在手術之前，我們有告訴 LR，他腦中主控的半部已經生病，而且必須要被移除，雖然會有某些人格上的改變，以及某些記憶可能的喪失，但是會有另外一半來接掌。他可能是一定會擔心而且會受到煩擾－但是他可能不會將此認為是一個死刑。換句話說，LR 將會因著確定 R 會繼續活著，而得到足夠的慰藉。但是在經過分割以及移植的手術之後的 L，就會將他自己的毀壞視為是一種死亡，而且將會不滿於 R 繼續在另外一個軀體中活著。

這個實驗似乎又再度地指出，如果不是在邏輯上，至少在心理學上，實質身體上的連續性乃是一個重要的考量因素。

實驗 7。一個人在臨床死亡一小段時間後被復甦，蒙受了一些記憶的喪失以及人格的改變。

這個實驗實際上已經進行過好多次了。(97) 以通常的臨床醫學試驗（沒有呼吸，沒有心跳）來看，死亡是真實的，但是當然大部分的細胞都還是活著的，而且大部分的人會認為他並沒有（真正地）死亡，並且在此之後他的確還是同一個人。這個實驗的重要性僅是在於充當下面一些實驗的背景。

實驗8。一個人死了，而且躺在那裡好幾天沒有人照料，因而發生過了生物的死亡以及細胞的死亡。但是現在有一椿奇事發生了；有一架太空船從大角星系的一個行星飛來，載來了一個屬於較先進人種的超級外科醫生，他應用他的技藝將這個死了而且爛掉的人救活了，並且把他其它次要的病也治好了。

(It is not, of course, suggested that any such elder race exists; the experiment is purely hypothetical, but as far as we know today it is not impossible in principle.)

The implications are apt to shake us. If decay is to be regarded as just another disease, with a possibility of cure, then when may the body be considered truly dead? If "truly" dead be taken to mean "permanently" dead, then we may never know when we are in the presence of death, since the criterion is not what has already happened to the man, but what is going to happen to him in the (endless?) future.

Experiment 9. A man dies, and decays, and his components are scattered. But after a long time a super-being somehow collects his atoms and reassembles them, and the man is recreated.

Once more, the difficulty or even impossibility of the experiment is not important. We also disregard the question of the possibility of identifying individual elementary particles. Is it the "same" man, in spite of the sharp physical discontinuity in time? If memory, personality, and physical substance are all the same, perhaps most of us would think so, even though we are disturbed by the black gulf of death intervening. But if we so admit, we must open the door even wider.

Experiment 10. We repeat the previous experiment, but with a less faithful reproduction, involving perhaps only some of the original atoms and only a moderately good copy. Is it still the same man?

Again, perhaps, we wonder if there is really any such thing as an individual in any clear-cut and fundamental sense.

（這當然不是意味著有任何這種較先進的人種存在；這個實驗純粹屬於假設性的，但是就我們當今所知，在理論上這並非是不可能的。）

這當中的意涵是輕易地就會震驚我們的。假使腐爛是被認為僅是另外一種疾病，而且有可能被治癒，那麼身體到什麼程度，才會真正地被認為是死亡呢？假使"真正地"死去，被認為就是意味著"永遠地"死去，那麼我們可能就永遠不知道麼時候我們是在死亡當中，因為其標準乃不是在於已經發生在一個人上面的，而是在於在（無窮的？）未來才要發生在他身上的。

實驗 9。一個人死了，而且爛了，並且他的成分散落各處。但是經過一段漫長的時光之後，有一個超級族類將其原子收集起來，把它們重新組合，而再度創造出這個人。

再講一次，有關這個實驗的困難，甚至於其不可能性都不重要。我們也忽略了有關確認個體基本粒子可能性的問題。就算不管明顯地在時間上實質的不連續性，到底這是不是"同一"個人呢？雖然我們會被因死亡的介入所產生的黯黑鴻溝所煩擾，但是假使其記憶，人格和實質的物質都是一樣的，或許我們大多數都會認為是同一個人。如果一旦我們已經如此地接納的話，那我們一定要將此大門開得更開一點。

實驗 10。我們重複前一個實驗，但是其再造的忠實度則較低，所包含的可能僅有一些原來的原子，而且所產出的僅是一個還算好的拷貝。那到底這樣還是不是同一個人呢？

因此再度地，我們或許還是會懷疑，在這麼乾淨俐落的以及根本的概念上，真的會有個人這種東西存在嗎？

Experiment 11. We repeat experiment 10, making a moderately good reconstruction of a man, but this time without trying to use salvaged material.

Now, according to the generally accepted interpretation of quantum theory, there is in principle as well as in practice no way to "tag" individual particles, e.g. the atoms or molecules of a man's brain; equivalent particles are completely indistinguishable, and in general it does not even make sense to ask whether the atoms of the reconstructed body are the "same" atoms that were in the original body. Those unfamiliar with the theory, who find this notion hard to stomach, may consult any of the standard texts.

If we accept this view, then a test of individuality becomes still more difficult, because the criteria of identity of material substance and continuity of material substance become difficult or impossible to apply.

Experiment 12. We discover how to grow or to construct functional replicas of the parts of the brain - possibly biological in nature, possibly mechanical, but at any rate distinguishable from natural units by special tests, although not distinguishable in function. The units might be cells, or they might be larger or smaller components. Now we operate on our subject from time to time, in each operation substituting some artificial brain parts for the natural ones. The subject notices no change in himself, yet when the experiment is finally over, we have in effect a "robot"!

Does the "robot" have the same identity as the original man?

Experiment 13. We perform the same experiment as 12, but more quickly.

實驗 11。我們重複實驗 10,製造出一個還不錯的再造人,但是這一次沒有去利用到回收的材料。

現在依據一般所接受的量子理論的詮釋,在理論上以及在實務上,並沒有一種方法可以來"標誌"單獨的粒子,也就是指一個人頭腦中的原子或是分子;同等的粒子之間是完全無法分辨彼此的,而且一般而言,甚至於去追問在再造的身體中的那些原子是否和在本來身體中"一樣的"原子,根本就是沒有意義的。那些不熟悉這個理論而覺得對此觀念感到反胃的人,可能可以去參閱任何一本標準的教科書。

假使我們接受了這個觀點,那麼對個體性的一種試驗,就會變得更加地困難,因為對物質性東西的身分和這些物質性東西的連續性,將會變成非常困難,甚或是不可能施加一套標準。

實驗 12。我們發現如何來培養出或是製造出腦部部分功能的複製品——可這能是生物性的本質,也可能是機械性的,雖然在其功能上是無法分辨的,但是藉著特殊的試驗,無論如何,是可以和天然的單元分辨出來的。這些單元可能是一些細胞,它們或者也可能是一些較大型或是較小型的部件。現在我們已經時常在我們的主體上進行手術,在每一次手術中都有用到某些人造的頭腦部件,來替代一些天然的部分。這個主體都沒有感覺到在他自己之中有任何變化,但是當整個實驗最後完成之後,實質上我們所獲得的就是一個"機器人"!

這個"機器人"和本來的人是否有相同的身分呢?

實驗13。我們進行和實驗12一樣的實驗,但是其速度要來得快許多。

In a single, long operation, we keep replacing natural brain components with artificial ones (and the rest of the body likewise) until all the original bodily material is in the garbage disposal, and a "robot" lies on the operating table, an artificial man whose memories and personality closely duplicate those of the original.

Perhaps some would feel the "robot" was indeed the man, basing the identity in the continuity, on the fact that there was never a sharp dividing line in time where one could say man ended and robot began. Others, well steeped in democracy and willing to apply political principles to biology, might think the robot was not the man, and ceased to be the man when half the material was artificial.

The subject himself, before the operation, would probably regard it as a death sentence. And yet this seems odd, since there is so little real difference between experiments 13 and 12; 13 merely speeds things up. Perhaps sufficient persuasion could convince the subject that the operation did not represent death; he might even be made to prefer a single operation to the nuisance of a series of operations.

Experiment 14. We assume, as in the previous two experiments, that we can make synthetic body and brain components. We also assume that somehow we can make sufficiently accurate nondestructive analyses of individuals. We proceed to analyze a subject, and then build a replica or twin of him, complete with memories.

Does the identity of our subject now belong equally to the "robot" twin?

在一個單一一次的冗長手術中，我們持續不斷地用一些人造部件來替代自然的腦部部件（而且身體其它部分也如法泡製），直到所有本來屬於身體的物質都進入了垃圾處理機，於是一個"機器人"出現了，平躺在手術台上的是一個人造的人，其記憶和人格乃是非常接近本尊的複製。

根據身分上的連續性，可能有些人會覺得這個"機器人"的確就是這個人，因為事實上，在時間上不曾有一個明確的分界線，讓一個人可以判定這個人的終結點和機器人的起始點。其他的人由於受到民主的深刻影響，並且願意將政治的原則應用在生物學上，可能就會認為這個機器人不再是這個人，而且當有一半的物質是人造的時候，就已經不是這個人了。

在進行手術之前，這個主體自己可能會將這認為是一種死亡的宣判。而這看起來似乎有點奇怪，因為在實驗 13 和 12 之間其實並沒有什麼實質上的差異；實驗 13 僅僅是將一切加速而已。或許如果有足夠的說辭，可以說服這個主體說這個手術並不代表著死亡；他甚至於有可能被改變成較喜歡這種一氣呵成的手術，而較不喜歡一序列手術的麻煩。

實驗 14。假設在前面兩個實驗中，我們可以製造出人造的身體和頭腦的部件。我們也假設，藉著某種方法我們可以對個體進行足夠精確的非破壞性分析。於是我們對一個主體進行了分析，並且接著製造出一個他的複製品或是雙生體，並且附帶著所有的記憶。

那麼現在我們所在實驗的主體的身分，是否相等地屬於這個雙生的"機器人"呢？

It might seem absurd to say so, but compare the previous experiment. There is scarcely any difference, especially since in experiment 13 the subject was under anesthesia during the operation; experiment 13 was virtually equivalent to destroying the subject, then building a robot twin. The only real difference between experiments 13 and 14 is that in experiment 14 both the original and the duplicate survive.

Experiments 15, 16, and 17. We repeat experiments 12, 13, and 14 respectively, but instead of using artificial parts we use ordinary biological material, perhaps obtained by culturing the subject's own cells and conditioning the resultant units appropriately. Does this make any difference?

In logic, one would think perhaps not, but blood is thicker than water. Some people might make a different decision on 15 and 16 than on 12 and 13.

Experiment 18. We assume the truth of an assertion sometimes heard, viz., that in certain types of surgery a patient under certain types of anesthesia suffers pain, although he does not awaken and afterwards does not remember the pain. The experiment consists in performing such an operation.

Most of us do not fear such operations, because we remember no pain in previous experiences, and because authoritative persons assure us we need not worry. Even a warning that the pain under anesthesia is real is unlikely to disturb us much, if we are not of very nervous temperament. Still less do we fear ordinary deep anesthesia, in which there seems to be no pain on any level, even though for the conscious mind this gulf is like that of death. Yet a child, or a person of morbid imagination, might be intensely frightened by these prospects.

這說起來似乎有點荒謬，但是如果和前一個實驗比較，其間其實幾乎沒有任何的不同，尤其是既然在實驗 13 中，這個主體在手術中乃是被麻醉的；實驗 13 實質上就等同於將此主體破壞掉，然後再製造出一個雙生體機器人。實驗 13 和實驗 14 間唯一真正的差異乃是在於，在實驗 14 中原本的人和其複製體都存活下來了。

實驗 15、16 和 17。我們分別重複實驗 12，13 和 14，但是我們不用人工的部件，而是用一般生物性的材料，或許是藉著培殖主體自己的細胞所獲得的，並且將所獲取的單元經過恰當地調理。如此是否就會有所差別呢？

在邏輯上，有人可能會認為是沒有的，但是畢竟血還是濃於水。有些人可能就會做不同的決定，要用 15 和 16，而不要 12 和 13。

實驗 18。我們假設有些時候所聽到的一個假說是真實的，那就是在使用某種類型麻醉下的某些類型的手術中，一個患者還是會有痛苦，只是他並沒有因此而醒過來，而且之後他也不記得這個痛苦。這個實驗就是包含了類似這種手術的進行。

我們大多數都不怕這種手術，因為在先前的一些經驗中，我們都不會記得有任何的痛苦，而且也因為一些專業權威，都保證我們不須要去擔心。假使我們不是那種緊張過度個性的人，那麼就算有警告說，在麻醉之下這個痛苦還是真實的，也不可能會多困擾我們。我們更不會懼怕那些一般深度的麻醉了，因為其中似乎沒有任何程度的痛苦，雖然對心理知覺而言，這個鴻溝就如同死亡一樣。但是一個小孩，或是對一個有病態幻想的人，就可能會被這種景象嚇得魂不附體。

Thus again we note a possible discrepancy between the logical and the psychological.

Experiment 19. A Moslem warrior is persuaded to give his life joyfully in a "holy war," convinced that the moment his throat is cut he will awaken in Paradise to be entertained by houris.

We draw the obvious but useful conclusion that, from the standpoint of present serenity, it is merely the prospect of immortality that is important.

Experiment 20. We pull out all the stops, and assume we can make a synthetic chemical electronic mechanical brain which can, among other things, duplicate all the functions of a particular human brain, and possesses the same personality and memory as the human brain. We also assume that there is complete but controlled interconnection between the human brain and the machine brain: that is, we can, at will, remove any segments or functions of the human brain from the joint circuit and replace them by machine components, or vice versa.

In a schematic sense, then, we envisage each of the two brains, the biological one and the mechanical one, as an electronic circuit spread out on a huge "bread board" with complete accessibility. From the two sets of components, by plugging in suitable leads, we can patch together a single functioning unit, the bypassed elements simply lying dormant.

To make the picture simpler and more dramatic, let us also assume the connections require only something like radio communication, and not a physically cumbersome coupling.

因此再度地我們又可以看到，在邏輯上和在心理上之間，可能存在著一種差異。

實驗 19。有一個回教勇士被說服來在一次"聖戰"中快樂地獻上其生命，相信當他的喉嚨被切斷的時刻，他也將在天堂中醒過來，而被一些美麗的仙女所照料。

從當今冷靜沉著的觀點來看，我們從中所獲得既明顯又有用的結論乃是，這其中僅有永生不死的期盼才是其重點。

實驗 20。我們克服了所有的阻礙，並且假設我們可以製造出一個人工的化學電子機械大腦，藉此除了可以有其他許多作用之外，還可以複製出某一特定人士腦部的所有功能，並且可以獲得和這個人一樣的人格特質和大腦的記憶。同時我們也假設，在人體大腦和這個機器大腦之間，有完整的，然而是可以被控制的相互連結：也就是說，我們可以任意地從電路連接點，來移除任何一些片段或是一些功能，而用一些機械性的零件來替代，同時也可以反向操作。

因此從解說的觀點來看，我們可以想像到會有兩個分別的大腦，也就是生物性的和機械性的，像是一個電子電路在一塊巨大的電路板上攤展開來，而且完全是可以隨時介入的。從這兩套零組件上，藉著插入恰當的端子，我們就可以將一個單一功能的單元併湊在一起，而被繞道的一些單元則就僅會處於休眠的狀態。

為了要使整個景象看起來較單純而且較明顯一點，就讓我們也假設，這些接點僅僅需要某些類似無線電波的通訊，而不需要一些麻煩的實體連結。

We might begin the experiment with the man fully conscious and independent, and the machine brain disconnected and fully dormant. But now we gradually begin disconnecting nerve cells or larger units in the man's brain, simultaneously switching in the corresponding units of the machine. The subject notices no change - yet when the process is completed, we "really" have a machine brain controlling a "zombie" human body!

The machine also has its own sensory organs and effectors. If we now cut off the man's sensory nerves and motor leads and simultaneously activate those of the machine, the first subjective change will occur, namely, an eerie transportation of the senses from one body to another, from the man's to the machine's. This might be enjoyable: perhaps the machine's sense organs are more versatile than the man's, with vision in the infra-red and other improvements, and the common personality might feel wonderful and even prefer to "live" in the machine.

At this stage, remember, the man is entirely dormant, brain and body, and the outside observer may be inclined to think he is looking at an unconscious man and a conscious machine, the machine suffering from the curious delusion that it is a man controlling a machine.

Next, we reactivate the components of the man's brain, either gradually or suddenly, simultaneously cutting off those of the machine, but leaving the machine's sensors plugged in and the sensors of the human body disconnected. The subject notices no change, but we now have a human brain using mechanical senses, by remote control. (We disregard such details as the ability of the human optical center to cope with infrared vision, and the duplication of the new memories.)

我們可以用一個意識完全清楚而且有獨立行為能力的人，而且在這個機器大腦還沒有連結上，還處於完全休眠之下來開啟這個實驗。但是接著下來，我們就可以逐漸地開始將人腦中的神經細胞或是較大的單元斷連，同時將機器中相對的單元連上。在這個受試體上，雖然看不到改變，但是一旦這整個試程完成之後，我們就"真正地"有一台控制著一個人體"殭屍"的機器腦了。

　　這台機器也有它自己的感應器官和反應器。假使我們現在就將人體中的感應神經和運動連線切斷，而且同時將機器中的這些功能連上，第一個主觀的改變就會出現了，那就是會有一種從一個軀體到另外一個軀體的怪異感應轉輸，也就是從人的傳送到機器的。這可能倒是一種非常的享受：機器的感應器官可能會比人體的要來得更為多采多姿，其視覺可能是紅外線的，並且還有一些其它方面的改良，因而一般的人可能會感覺得很不錯，甚至於會寧願"活在"這個機器裡面。

　　請記得在這個階段，這個人，不管是大腦或是身體，是處於完全休眠的狀態，而且外在的觀察者可能會傾向認為，他所看到的乃是一個沒有意識的人和一台有意識的機器，而這台機器則是蒙受一種怪異的幻覺，好像是有一個人在控制一台機器。

　　下一步，我們重新啟動人腦的部件，這可以是逐漸的也可以是突然的，同時將機器的部分切斷，但是將機器的感應器保持接通，而把人體的感應器斷連。這個受試體雖然看不到改變，但是現在我們就有一個人腦，透過遙控在使用機器的感官。（至於人體視覺中心如何應付紅外線的影像以及新記憶的複製的能力，我們先將其細節忽略。）

Finally, we switch the human effectors and sensors back in, leaving the man once more in his natural state and the machine quiescent.

If we perform this sort of exchange many times, the subject may become accustomed to it, and may even prefer to "inhabit" the machine. He may even view with equanimity the prospect of remaining permanently "in" the machine and having his original body destroyed. This may not prove anything, but it suggests once more that individuality is an illusion.

Discussion and Conclusion. In discussing these hypothetical experiments we have touched on various possible criteria of individuality - identity of material substance, continuity of material substance, identity of personality and memory, continuity of personality and memory - and seen that none of these is wholly satisfactory. At any rate, none of these, nor any combination, is both necessary and sufficient to prove identity.

One cannot absolutely rule out the possibility that we have missed the nub of the matter, which may lie in some so far intangible essence or soul. However, such a notion seems inconsistent with the ease with which man can instigate, modify, and perhaps actually create life, and with several of our experiments.

The simplest conclusion is that there is really no such thing as individuality in any profound sense. The difficulty arises from our efforts first to abstract generalities from the physical world, and then to regard the abstractions, rather than the world, as the basic reality. A rough analogy will help drive home the point:

最後我們將人體的感應器和反應器都再接回去，讓這個人再度回歸到其自然的狀態，而讓機器回歸靜止。

假使我們進行這類的交換動作許多次之後，這個受試體可能就會習以為常，甚至於可能會較喜歡"住在"這台機器裡面。他甚至於還可能會對永遠處在這個機器"裡面"，並且將其本來的軀體摧毀的可能，以平靜平常的心看待。這樣雖然不能證明任何事情，但是這卻再度地指出個體性可能只是一種幻象。

討論和結論。在討論這些假設性的試驗中，我們已經有觸及個體性中的各種可能的基準條件－物質實體的身分，物質實體的連續性，人格和記憶的身分，人格和記憶的連續性－而且也看到其中沒有一個是能夠完全充分自足的。不管如何，其中沒有一種，甚或任何種類的結合，在驗證身分的條件上，既是必要而且是充分的。

沒有一個人可以絕對地排除我們沒有去掌握到這件事情的要點的可能性，而其關鍵點可能就在於一些極度難以衡量的要素或是靈魂。然而，這種觀念看起來似乎是和人可以很容易地來啟動，修飾，甚或於真正地創造出生命，以及和我們許多的試驗結果是不一致的。

最簡單的結論乃是，以任何深度的觀點看來，其實是沒有一種所謂個體性的東西。這種窘境乃是起因於我們開始時，努力地要從實體的世界中擷取一些共通性，但是接著我們卻將所擷取的，而不是這個世界本身，誤認為就是基本真理。用一個粗淺的類比來幫助我們點出其中的真義：

The classification "man" is useful, but not sharply definable. Is a freak a man? Is an aborted fetus a man? Is a pre-Neanderthal or other "missing link" a man? Is a corpse a man if some of the cells are still alive? And so on. A label is handy, but objects may be tagged arbitrarily. In the physical world there is no definite collection of objects which can be called "men," but only shifting assemblages of atoms organized in various ways, some of which we may choose to lump together for convenience. Let us then cut the Gordian knot by recognizing that identity, like morality, is man-made and relative, rather than natural and absolute. Identity, like beauty, is partly in the eye of the beholder. It is only partly existent, and partly invented. Instead of having identity, we have degrees of identity, measured by some criteria suitable to the purpose.

The result is wonderful: we have lost our souls, but gained heaven, in a certain sense. Perhaps few of us, even if intellectually convinced that identity is an illusion and death therefore unimportant, may be able to translate this into emotional acceptance, or will want to. But we can now persuade ourselves that death need never be regarded as absolutely final - since it is always possible, at some distance in space, time, and matter, for reasonably close duplication or resuscitation to occur - that is, for physical reincarnation, with memory or without. This possibility can dull the edge of desperation for those unable to obtain first-class freezer accommodations for themselves or their families.

"人"這個分類是有用的，但是卻是很難被精確地定義。一個畸形人是人嗎？一個被墮胎的胎兒是人嗎？一個先尼安德塔人或是其他的"遺失的環節"是人嗎？一具屍體，假使其中的一些細胞還是活著的，是人嗎？還有好多例子可舉。貼上標籤是輕而易舉的，但是有些東西就可能會被亂貼標籤。在實體的世界中，沒有一個確切的物件的集合可以被稱為是"人"，而是只有原子的變化排列，以各種不同方式的組合，而為了方便起見，我們可能會將其中的某些東西送作堆。因此讓我們把那難以解開的糾纏剪斷，例如將道德品行的身分辨識，認為是人造的而且是相對的，而不是自然的和絕對的。而例如對美麗的身分辨識，其中部分則是由於情人眼中出西施。它僅有部分是存在的，而另一部分則是被發明出來的。因此，與其用一個身份辨識，我們不如來用一個適合目的的某些基準所量度出來的身份辨識的梯級。

　　如此，所獲得的結果絕佳：從某些觀點看來，我們雖然喪失了靈魂，但是我們卻贏得了天堂。甚至於我們當中有一些人，在智理上都信服身分僅是一種夢幻，而認為死亡是不重要的人，或許可能會將此轉變成感性上的接納，而願意要如此行。但是因為在某一空間，時間和物質的遠方，總是有可能來讓相當完美的複製或是復活發生，也就是有記憶或是無記憶的肉體輪迴，所以現在我們都可以說服自己，死亡不需要被認為是一個絕對的終結。對那些無法為他自己或是他的家人找到頭等的冰凍器來容身的人，這種可能性將可以消除其煎熬的苦楚。

CHAPTER IX

The Uses of Immortality

永生的利用

CHAPTER IX

The Uses of Immortality

When people blithely assert that they "wouldn't want to live forever," it usually means only that they have not really thought about it.

The disclaimers fall perhaps into two main categories. The first concerns alleged moral considerations. "When I'm called, I'm ready to go . . . We have to step aside for our children. We shouldn't impose ourselves on posterity . . . Trying to hang on beyond our natural span is undignified and cowardly . . . Birth, growth, and death form a natural and necessary cycle. . . Fear of death is a sign of immaturity."

As far as "imposing ourselves on posterity" is concerned, some remarks have already been made, and a good deal more will be said in the last chapter. At this point it may be well just to insert a reminder that the freezer program should make everybody less desperate and the world more stable; it may even swing the balance in preventing a nuclear war. If this is true, then without the freezer program there may be no posterity. In that case, our descendants need it as much as we do.

When someone continues to insist that he would not want unlimited life, he is often wearing a mask which can be lifted by putting the question just a little differently.

第九章

永生的利用

當人們輕率地表示說他們"不會想要去永遠活著"，這通常都僅僅意味著他們沒有真正地去思考過它。

這些棄權的人或許可以歸納成兩大類。第一類的人擔憂所謂道德上的一些考量。"當我要被召回時，我就會準備要走……我們必須要為我們的子孫留下生存餘地。我們不能把我們自己強壓在子孫上面……想要在我們自然的收命隻外苟延殘喘，不僅是沒有尊嚴而且是懦夫的行為……誕生，成長，和死亡構築成一個自然而且是必要的循環……對死亡的恐懼乃是一種不成熟的表徵。"

針對"把我們自己強壓在子孫上面"而言，我們已經有做某些註解，而在本書最後一章也將會有更多的闡述。在這個時點，或許應該要插入一個提醒，那就是有了冰凍計劃後，應該可以讓每一個人的懊惱減少，而且世界會更穩定；它甚至於可以在防止一個核子戰爭上，調動其平衡點。假設這是真的，那麼如果沒有了冰凍計劃，就有可能就不會有後代子孫。在這種狀況下，我們的子孫和我們一樣地都需要它。

當有某些人一直都堅持說他不想要無限的生命時，他是一個時常戴著面具的人，如果用一種稍微不同的問話方式，就可以將其面具拿掉。

Let us ask first: "In case of a severe infection, would you refuse penicillin to stave off the 'natural' conclusion of death?" He can hardly claim he would. We ask next: "If a serum were to appear on the market, guaranteed to add twenty vigorous years to your life, would you refuse it?" He is not likely to say he would, nor would he refuse a perfected immortality serum.

And now we see his true face: he wants immortality, all right, but he wants it on a silver platter. It is not life he objects to, but effort and risk. Far from being stoic, or resigned, or well-adjusted, or complacent, or mature, or philosophic, or self effacing, or altruistic, or any of the other dignified things
he pretends to be, he is merely myopic and nervous.

A slightly different way to unmask such a person is simply ask him what span of life he would pick, if he could make his own choice merely by wishing. Would he pick exactly his "natural" span, neither more nor less? Would he willingly accommodate himself to whatever accident or ailment would do him in, in the ordinary course of events, seeking neither to shorten nor prolong his life? Merely to ask these questions reveals the absurdity of affirmative answers.

The notion that superior personalities accept death more readily has also been denied, for example, by Dr. C. Knight Aldrich, chairman of the department of psychiatry at the University of Chicago. He writes:

讓我們首先問道："在一個嚴重感染的狀況下，你是否會拒絕使用盤尼西林來遏止'自然的'死亡定論呢？"他幾乎不可能宣稱說他會如此。接著我們再問："假使有一種血清在市場上出現，保證可以增加你二十年活力旺盛的壽命，你會拒絕它嗎？"他是不可能說他會的，而且他也不會去拒用一種技術成熟的不死血清的。

　　因此現在我們就識破他的真實面貌了：沒錯，他是想要不死的，只是他所要的是烹調完善的東西。他所反對的不是生命本身，而是必須負擔的代價和承受的風險。他絕對不像是他所假裝的禁慾，或是認命，或是圓融，或是滿足，或是成熟，或是哲思，或是無我，或是利他，或是其它一些尊貴的東西，其實他只是近視的而且是神經質的。

　　另外一種稍微不同的方式來剝下這樣一種人的面具，就是直接問他，假使他可以只要想活多久就可以活多久，那他會想選擇多長的壽命。那麼他會剛好選擇他的"自然的"壽命，不多也不少嗎？他自己是否會心甘情願地承受在日常生活事件中，任何意外或是疾病會對他造成的影響，而不會去尋求縮短或是延長他的生命呢？只要去問一問這些問題，就可以讓回答說是的荒謬性流露無遺。

　　有關有較優越人格的人會較能夠接受死亡的觀念，也已經被例如芝加哥大學精神病學系的系主任C. Knight Aldrich博士所否定了。他寫道：

"However, my experience, both clinical and extra-clinical, suggests that it is particularly difficult for the strong, well-integrated personality to accept with equanimity the idea of his own death.

"Strength of personality may help a patient not so much to void depression in anticipation of death as to conceal depression in others. On the other hand, many patients who seem to accept death with real equanimity have depressions that antedate their fatal illnesses or they have lost interest in living as a result of pain or disability. Their apparent realistic courage indicates that they have given up life and are welcoming death. Death can be faced more readily if there is little to lose by leaving life than if there is a great deal to lose." (1)

A few of my friends have expressed fear that the cowards will embrace the freezer program most eagerly, while the brave and the dignified may spurn it. This notion seems to be negated, not only by the theoretical considerations already put forward, but also by observation. Among the many who have talked and written to me, it is by and large the weak and timid who hang back - is it not the nature of the weak and timid to hang back? - while the strong and hold spirits usually seize the concept with delight. Only those embrace death who are half dead already. The ones who surrender are those who are already in retreat.

"不管如何，我的經驗，無論是醫學臨床上或是臨床外，都顯示出那些堅強，人格完美的人，對其自己死亡的意念，特別無法以平靜的心情接納。"

"在面對死亡的來臨，人格上的堅強不可能幫助一個病人多少來消除其鬱卒，最多只能在別人面前將其鬱卒隱藏。在另外一方面，許多看起來似乎可以以真正平靜的心情接受死亡的病人，在他們因疾病而致命之前都有憂鬱症，或都會因著痛苦或是身體的殘障，而導致對生命失卻興趣。他們表現出來的真實勇氣在在都顯示出他們已經放棄了生命，而迎迓著死亡。假使可以使生命的離開沒有任何的損失，比起要為死亡付出極大的代價，那麼死亡就可以較容易地面對。"（1）

我的一些朋友表示他們擔憂會有許多懦弱者會更汲營於擁抱這個冰凍計劃，而那些勇敢和有尊嚴的人則會排斥唾棄它。不僅是從已經提出的理論上的考量，以及從實質的觀察而言，這種意念似乎是無的放矢。在曾經和我交談過的人以及寫信給我過的人當中，大多數都是那些懦弱膽小的人會有所退縮——退縮不就是懦弱膽小者的天然本性嗎？——反而是那些勇敢者和堅毅不拔者，通常都會以喜悅的心情來把握這個概念。只有那些緊緊擁抱住死亡的，才是沒死而棺材已經進一半的人。那些投降者才是已經在撤退的人。

You, Better than New

Disclaimers in the second category question whether extended life is worthwhile, even to the individual. "I've had a full life. I'm already bored and couldn't endure a second life. I wouldn't like a futuristic world. . . . There would be nothing to do . . . I wouldn't fit in." Etc.

The main difficulty is that few people have the remotest conception of what the future will he like; they think of it dimly as mid-twentieth century, plus maybe sliding sidewalks, family helicopters, and a twenty-hour work week. They fail to understand that the differences will be qualitative as well as quantitative.

In particular, they completely fail to grasp that people will be different, including themselves. Mental qualities, including both intellectual power and personality or character, will be profoundly altered, not only in our descendants but in ourselves, in you and me, the resuscitees.

That genetic science will enable us, sooner or later, to mould our children as we please seems almost taken for granted by the experts and by all laymen with a spark of imagination. We might for example quote Dr. Philip Siekevitz:

比新的還好的你

第二類的放棄者會質疑說延長的生命是否質得,甚至於針對其個人而言。"我已經活夠了一輩子了。我已經厭煩了,而且不可能再去忍受一個第二度的生命。我不會去喜歡一個太未來性的世界⋯⋯將會沒有啥事可做⋯⋯我會難以適應的⋯"等等。

其中最大的障礙乃是在於未來會像是什麼境界,大部分的人都一點概念也沒有;他們把它模模糊糊地想像成二十世紀的中期一樣,或許在添加一些自動滑行的人行道,自家用的直昇機,以及每週只要二十四小時的工作。他們都無法去理解到其改變不僅會是在數量上,而且是會在質量上的。

尤其是他們根本就沒有去完全掌握到人們將會不同,連他們自己也是一樣。心理上的品質,其中包括了心智的能力以及人格或是特質將會有深度的轉變,而且不僅在我們的子孫上面,也在所有我們上面,包括你和我這些被復甦的人。

基因科學遲早將會讓我們能夠按照自己的喜好來塑造我們的子代,這乃是專家們和所有具有一點想像力火花的所有業餘人士認為幾乎是理所當然的事。譬如我們也可以引用Philip Siekevitz博士的一段話:

"For I think we are approaching the greatest event in human history, even in the history of life on this earth, and that is the deliberate changing by man of many of the biological processes. ... Already we can very easily produce mutants in bacterial strains; we will soon be able to control these changes; and it is not such a big jump from bacteria to plant, to animals, or to man himself . . . we will be able to plan ahead so that our children will be what we would like them to be - physically and even mentally." (105)

Some of us may suspect that from uncontrolled mutations in bacteria to controlled mutations in man really is a pretty big jump, and it may take quite a while. But time, fortunately, is what we have plenty of. Sooner or later, these achievements will be realized. Professor Hermann J. Muller, Nobel Prize winner in genetics, has said: "I am convinced that he [man] will remake himself [genetically] . . . we may attain to modes of thought and living that today would seem inconceivably god-like." (76)

Such breath-taking predictions are readily accepted by many intelligent laymen - but only as interesting speculations of no direct concern. These notions take on entirely different colors when we realize that we, personally, may be there to witness these titanic events - and that we will have to deal with these supermen.

What will they be like, these genetically planned and engineered descendants of ours?

"因為我認為我們已經在接近人類歷史中的,甚至於是在地球上生命的歷史中的一個最大的事件,而這就是經由人可以來任意地改變許多生物的程序．．．．．．．我們線再已經很容易地就可以製造出許多細菌菌種的突變種;我們不久將有辦法來控制這些改變;而從細菌到植物,到動物,或是到人類本身,並非是一個極大的跳躍．．．．．．．．我們將可以來預先規劃,使得我們的子代變成是我們所想要的樣子──不管是在身體上,甚至於是在心理上。"(**105**)

我們其中有些人可能會認為從在細菌中沒有受控制的突變,到在人體上受控制的突變,其實是一個相當大的跳躍,而這可能會耗時許久。但是很慶幸地,時間,我們有的是。遲早這些成果將會一一被實現的。在基因學上諾貝爾獎得主 **Hermann J. Muller** 教授說道:"我深信他(人類)將會以基因學的方法改造他自己．．．．．．我們可能到達連當今的人都會認為像是一則不可思議地神話一般的一種思維模式和生活型態。"(**76**)

這些動人心弦的預言是許多有識之士很容易就可以接納的──但是僅是出於好玩的猜測而已,而沒有直接切身的關係。當我們發覺到我們本身可能要親自當場去驗證這些震天動地的事件時,這些意念就會完全地豬羊變色──而且我們將必須要去面對這些超人。

這些經過基因科學規劃過而且設計過的我們的子孫,他們將會像什麼呢?

On the physical side, they will be strong, handsome, and healthy, but beyond this it is impossible yet to say. They may look almost exactly like people today, or they may not; certainly the current human design leaves plenty of room for improvement.

For example, Professor Muller has pointed out the absurdity a multi-purpose mouth. "[An alien] would find it most remarkable that we had an organ combining the requirements of breathing, ingesting, tasting, chewing, biting, and on occasion fighting, helping to thread needles, yelling, whistling, lecturing, and grimacing. He might well have separate organs for all these purposes, located in diverse parts of his body, and would consider as awkward and primitive our imperfect separation of these functions." (78)

While Dr. Muller may have stretched it a little -- not all of us want to separate our yelling from our lecturing, and a special organ to thread needles will find few buyers even in a ladies' sewing circle-the point is surely well taken. Imagine the incalculable benefit in teen-age happiness alone, if one could eat, chew gum, and talk on the telephone, all at the same time - without the danger of strangling!

But the great changes will be those in intellect and personality. And if our descendants are all super-duper whiz kids, even if they are kind and good, how can we compete? How can we live? The problem is real, but there are solutions.

就身體面而言，他們將會是強壯的，英俊的，而且是健康的，但是超過這個目前就不可能知道了。他們看起來大概可能會和當今的人們一樣，他們可能也會不一樣；當然時至目前人體的設計實在是還留有許多可以改進的空間。

譬如說，**Muller** 教授就有指出一個多重功能嘴巴的荒謬性。"（一個外星人）可能會發覺我們竟然有一個器官，將呼吸，攝食，嚐味，咀嚼，吃咬，以及偶而的打架，幫助穿針線，喊叫，吹口哨，說教，和扮鬼臉的需要全部混在一起，而感到非常地訝異。他大可能會有不同的器官來負責這些不同的功能，分布在他身體各個不同的地方，而會認為我們這些功能不完美的分工是齷齪的，而且是原始的。"（**78**）

雖然 **Muller** 博士可能是有點瞎掰──不是所有的人都會想要去將喊叫和說教的功能分哩，而且有一個特別的器官來穿針線，甚至於在有用真線的婦女圈子中，將也找不到幾個同意者──但是其重點的確是可以好好採納的。假使一個人可以吃，嚼口香糖，並且講電話，全部在同一個時間，而且不會有窒息的危險，光就想像其在青少年人的快樂上就可產生不可計量的效益!

然而最大的改變將還是會是那些在智能上和人格上的。而假使我們的子孫們都是超級頂好的精明小孩，甚至於假使他們也都是既善良又乖巧，那麼我們如何能夠來競爭呢？這個問題是切實的，但是是有解決方案的。

One solution, of course, is to refuse to breed supermen. Such issues will be hotly debated in all the parliaments of earth, with unpredictable results. But probably the issue will fade, for several reasons.

In the first place, we will not necessarily be resuscitated the moment it becomes technically feasible, unless we insist. If it should happen that at this time genetic improvement is far advanced but individuals cannot yet be much improved, then resuscitation may be delayed, conceivably.

Second, even if for a time we have to live with superior descendants, a modus vivendi may be found. There will be means for reducing envy on the one hand and arrogance on the other, and for enabling the individual to enjoy what he has; more will be said about this shortly. It is also important to remember that no disadvantage need be permanent; we can no doubt summon considerable patience when we know that we only have to wait a while, and science will improve us further.

Third, we shall probably be supermen ourselves shortly after resuscitation. Somatic improvement may stay abreast of, or get ahead of, genetic improvement. Designing a new model will not necessarily be easier than overhauling the old. It should become possible to perform extensive improvement in living individuals by various biological techniques, for example, using regeneration together with somatic mutation, microsurgery, and psychosurgery.

當然有一個方法就是拒絕去孕育超人。這個議題將會在地球上的所有立法單位中被熱烈地辯論，而且不會有可預期的結果。但是可能這個議題將會自行地退色，其中有許多個理由。

　　首先，我們將不一定在技術上變成可行的當下就要來被復甦，除非我們堅持要如此。假使碰巧在這個時段，基因改善的技術是非常地先進，但是個人的人格還是沒有被改良多少，那麼復甦的動作可想像地就可以被暫緩。

　　第二，就算我們必須要和這些較優越的子孫一起生活一段時間，還是可以尋求出一種暫時的生存模式。其中一定有一些方法來一方面減少羨慕，另一方面來降低傲慢，而且來讓個人可以滿足於自己所有；很快地就會進一步討論這個。記得說沒有一種優勢一定會是持久永遠的也是非常地重要；當我們知道我們只要稍加等待，而且科技一定會進一步改良我們，無疑地我們就可以提振出相當的耐心。

　　第三，在我們被復甦不久之後，我們自己將有可能就是超人。軀體上的改良可能會和基因上的改良同步並行，甚或會領先。設計出一個新的型式，將不見得會比翻修舊的來得容易。利用各種生物上的技術，來在活著的個體上進行廣泛的改良應該會變成可行的，例如利用再生技術和體細胞變異，微手術技術和心理手術技術的並用。

Besides biological changes, there is also vast potential in the use of prostheses, mental as well as physical - for example, by coupling a human mind to an electronic computer. Some different suggestions, but in the same general direction, have been made by Dr. R. M. Page, Director of Research, U. S. Naval Research Laboratory, Washington; he envisages ultra-rapid communication between man and machine by a sort of electronic mind-reading, and thinks it might be achieved in fifty years. (85) Thus all the resources of a huge computer may some day be in the direct service of a man's mind; it might even be said to be part of his mind, when hooked in on either a temporary or permanent basis. The man-machine combination may well be far superior to any purely biological superman, in which case we shall be immediately equal to our descendants.

The best advice for success in life has always been to choose your parents wisely; and now, in effect, this will become possible. Collectively, if not individually, we can expect to design ourselves, selecting the desired traits and abilities. Of course, the alarmists will protest that there may be unforeseen consequences, and such presumption is dangerous. And we must agree; that it is. But we can only choose between dangers, and not escape them. Doing nothing also constitutes a choice, and often a poor one. (Stock market players often forget that every day they hold a stock they have, in effect, except for overhead, made a new decision to buy that stock in preference to all others.)

Living has always been dangerous; and now, for the first time, dying will be dangerous too. But most of us will prefer the danger of our activities after resuscitation to the safety of no activity at all.

除了生物性的改變之外，在利用心理和生理上的彌補技術也具有極大的潛力，例如，將一個人的心思和一台電腦的結合。美國華盛頓州，海軍實驗室的研究室處長 R. M. Page 博士也提出了一些不同的，但是方向大致是一樣的建議；他想像出人和機器之間，利用某一類的電子讀心術的一種超快速溝通，而他認為在五十年之內就可能會完成它。（85）因此一台巨大電腦的所有功能，有一天可能會直接用來服伺一個人的心思；當其被暫時性或是永久性地連接上時，它甚至於還可以說是他的心思的一部分。這種人—機器的混合可能會比任何純粹是生物性的超人還要優越上許多，在這種情形下，我們將會馬上就和我們的子孫平等。

要在人生中成功，最好的忠告往往就是要去選對你的父母；而現在實質上，這樣已經變成可能了。就集體而言，假使不談個體上的，我們將可以來設計我們自己，選擇所想要的特徵和能力。當然，那些警世主義者將會抗議說其中可能會有一些無法預見的後果，並且這種假設是危險的。而我們則必須要同意；真的就是這樣。然而我們只能在危險和不能解脫它們之間作一個抉擇。沒有作為也構成一種選擇，而其通常都是一個壞的。（股票市場的玩家常常會忘記了每天他手中已經持有的某一張股票，實際上，除了額外成本的支出，他還會作下一個新決定來買同樣一張，而不去買所有其它的股票。）

活著一直都是危險的；而現在破天荒地，死亡將也是危險的。但是我們當中大多數的人在復甦之後，將會選擇我們有所作為的風險，而不會選擇完全無所作為的安全。

Just what our activities may be in the world of the future is not easy to picture. At least for a considerable period, there will still be economically productive work in the form of scientific investigation, administration, education, and many kinds of artistic endeavor. There will be many activities involving human relations which, although perhaps not economically productive, will give that "needed" and "useful" feeling; for example, taking an interest in your children's or parents' troubles, or participating in politics. Certain simple pleasures are likely to wear quite well - things like exercising the muscles and glands, playing with the children or great-grandchildren, enjoying the lakes and forests.

This may seem rather thin at first. For instance, how many people can be painters, or writers, or composers, or sculptors? The answer is, maybe everybody! Nobody will be stupid - not by today's standards, and probably not by tomorrow's, since there is likely to be more homogeneity. There will simply be more artists and smaller average audiences. I'll buy your painting, and you'll buy my music; and each will enjoy doing his own work, and each will appreciate the other's.

If this still sounds unconvincing, we can help make our point with a wellknown and venerable story. A communist was exhorting an audience of laborers: "Comes the Revolution, you'll eat strawberries and cream." One worker objected, "But I don't like strawberries and cream." The agitator glared at him. "Comes the Revolution, you'll like strawberries and cream!"

在未來的世界中，究竟我們的活動會是什麼乃是不容易來描繪的。至少有一段相當長的時間，還是會有一些經濟上具有生產力的工作，像是科學研究，管理，教育，和許多類別的藝術活動。其中將會有許多和人際關係相關的活動，這些雖然在經濟上不具有生產力，但是卻可以給予那種"被需要"和"有用的"感覺；例如，對你的孩子們或是父母親的問題有所關懷，或是參與政治。有一些單純的樂趣有可能是會歷久不衰的——像是肌肉和腺體的運動，和兒女或是孫子玩耍，享受湖泊和森林。

這個剛開始時看起來似乎會有點薄弱。譬如說，會有多少人可以成為畫家，或是作家，或是作曲家，或是雕刻家呢？其答案是，可能是每一個人！沒有人會是愚蠢的——以當今的標準來看，而且以未來的標準也不是，因為均勻性是有可能會越來越高。未來就是會有越來越多的藝術家，而其平均的觀眾將會越來越少。我會來買你的畫作，而你會來買我做的音樂；而每一個人將會享受他自己的創作，每一個人並且會去欣賞別人的。

假使這樣聽起來還不夠說服力，我們可以用一個眾所皆知而且是值得尊敬的故事，來幫助增強我們的論點。有一個共產人士曾在對一群勞工聽眾慫恿說："革命成功之後，你就會有草莓和奶油吃。"一個工人就反抗說，"可是我又不喜歡吃草莓和奶油。"這個煽動者以閃爍的眼光瞪著他說，"革命成功之後，你就會喜歡吃草莓和奶油！"

The skeptics must be continually reminded that not only will the world be changed, but themselves also both their ability to perform and their capacity to appreciate and enjoy. There are already many forerunners of these developments.

There is now considerable use - sometimes excessive use - of tranquilizers on the one hand and "psychic energizers" on the other. Mood is known to be related to hormone and enzyme balance. Depression and anxiety can often be relieved by such drugs as epinephrine and adrenochrome, and many other drugs are known to affect personality. Furthermore, many common mental disorders may be at least partly chemical in nature; e.g., schizophrenia seems to be related to the production of a substance called taraxein. (43)

Some of the future potentialities have been indicated by Dr. A. Hoffer and Dr. R. Humphrey Osmond, Canadian psychiatric researchers: "Psychopharmacology may help us learn how to think clearly however distracting personal and other calls may be, without however preventing us from indulging, when we need it, in the boldest imaginings. Such capacities developed in an increasing number of our species would be as effective as a beneficial mutation and we think, far more easily achieved." (43)

Elsewhere, in discussing experiments with "psychedelic" or "mind-manifesting" drugs, Dr. Hoffer writes:

那些懷疑者一定要被一再地提醒，不僅未來的世界會被改變，還有他們自己，包括他們做事的能力以及他們欣賞和享受的範疇都會被改變。在這些方面的進展，目前已經有許多前驅人士。

目前在某一方面已經有相當程度的使用——有些時候是過度的使用——鎮定劑，在另一方面也有使用"精神增能劑"。大家都知道心情是和荷爾蒙以及酵素平衡有關的。憂鬱證和焦慮正常常是可以用類似腎上腺素和腎上腺素紅的一些藥物來舒緩，而且有許多其它的藥物已經被知道是會影響人格的。進一步而言，許多一般精神上的疾病，本質上至少在某一部分可能是化學性的；例如，精神分裂症似乎是和一種稱為 taraxein 物質的產生有關係的。(**43**)

加拿大精神醫學研究科學家，**A. Hoffer** 博士和 **R. Humpphrey Osmond** 博士就有指出一些未來的潛在可能："不管是精神散渙的或是其它的可能的原因，以最大膽的想像看來，當我們有所需要時，精神藥物學可能可以幫助我們學習如何來有清明的思緒，然而並沒有辦法防止我們上癮。在越來越多的人種上研發類似這樣的可能，將會和良性的突變是一樣地有效，而且我們認為其將會更容易達成。" (**43**)

在其他方面，有關探討"迷幻狀態"或是"心理展現"藥物試驗上，**Hoffer**博士寫道：

"Thought becomes creative, one's horizons are widened, and the world and its problems are seen with a fresh eye . . . Over half of our patients who achieve a psychedelic experience are subsequently much better people. For example, out of more than half of a series of sixty alcoholics treated in this way over one-half are now sober and good citizens and certainly much happier than they were before. Volunteers who have experienced this type of reaction find to their surprise and pleasure that they are more mature, more tolerant, and have a broader outlook on life." (42)

Some progress has been made in tracing motivation, as well as sensation, to local centers in the brain, according to Dr. James Olds, physiologist at U.C.L.A. It appears that rats can be made to gratify the drives of hunger, thirst, digestion, excretion, and sex by self-stimulation of their brains with electricity. (The rats were allowed to manipulate controls which turned on the current, the electrodes being inserted in suitable regions in the brain; the technique is called ESB, for Electronic Stimulation of the Brain.) The report says that "Some of the animals have been seen to stimulate themselves for twenty-four hours without rest and as often as 5,000 times an hour." (84)

This is in some ways an obscene experiment, with sinister overtones. (The same may be said of many biological experiments.) But it underscores the possibility of finding "happiness" partly by working on oneself, rather than by working on the environment.

"思想變成具創造力，一個人的水平線會被擴展，而對整個世界和其中所有的問題，都會以一個嶄新的眼光來看待........我們的病人中有超過一半的人，在經歷過迷幻的經驗之後，都會變成較健康的人。譬如說，對六十個酗酒的人進行一序列的這種治療，其中有超過一半的實驗中都有超過一半以上的人，現在都較清醒而且為人更好，並且確定地比以前都要來得快樂。曾經經驗過這類反應的志願受試者，都很訝異地而且快樂地發現他們變得更為成熟，更加具有耐心，而且對人生都有較廣闊的前瞻性。"（42）

依據美國加州大學洛杉磯分校的生理學家 James Olds 博士，在追溯動機，以及感性回到大腦中的一些局部性的中心點，已經有達到一些進展了。使用電流來自我刺激老鼠的腦部，對飢餓，口渴，消化，分泌和性慾的驅策，它們似乎可以被變成得到滿足。（這些老鼠被容許來操作控制點，可以來啟動電流，而一些電極則是插在其腦部中的一些適當區域；這種技術簡稱為 ESB，代表著腦部的電子刺激，Electronic Stimulation of the Brain。）這篇報告中指出"這些動物當中的某些隻有被看到會刺激它們自己長達二十四小時，片刻也不休息，而且次數高達每小時 5,000 次。"（84）

這個實驗在某些方面是有點不雅，而且帶有一些猥褻的意味。（有許多的生物實驗都可以說是如此的。）但是這點出了追求"快樂"部分是有可能從自身來的可能性，而不是從環境中求得的。

Professor Rostand has also emphasized the possibilities in improving individuals. ". . . intelligence . . . also character can be affected by chemical dosing . . . The future may bring the use of medicines that would favor social behavior, kindness and devotion . . . the possibility cannot be excluded that there may come into being a psychosurgery whose aim would be to raise the individual above himself . . ." (95)

I would go further than the cautious optimism of these scientists and say that, given enough time, these dazzling developments and many others have a high degree of probability.

But even if one grants all this, there remain the questions of long-range goals, of fundamental values and motivations, of the nature of happiness. If we dare face immortality, must we not also face the profoundest problems of man and the universe?

The Purpose of Life

It is possible we shall never be able to find "ultimate" values or "ultimate" goals. It is also possible that there are built-in conflicts or paradoxes in the human mind on the deepest level, so that in the end tragedy cannot be avoided. Not every problem has a solution.

At present, however, such speculations seem all but futile. We are too raw even to frame proper questions, let alone understand the answers. The very structure of our brains may need improvement before we can apprehend the secrets of the cosmos.

Rostand 教授也強調了改善個人的一些可能性。"………
智慧……還有個性是可以用化學藥物來影響的………未來可能
會引進一些會對社會行為，善良和奉獻有益的化學藥物………
不可以排除將來會有心理手術的可能性，其目的乃是用來將個
人提升而超越自己的………"（95）

越過這些科學家的審慎樂觀，我會進一步認為，只要假以
時日，這些以及許多其它耀人眼目的發展將會變成具有高度的
可能性。

但是就算這一切都成真了，還是會留下有關基本價值和動
機，以及快樂的本質的那些長程目標的問題。假使我們敢去面
對永生不死，我們不是也應該要去面對人和宇宙之間的那些最
深邃的問題嗎？

生命的目的

我們將永遠無法追尋到"最終的"價值或是"最終的"目標，
這是可能的。在人類心靈的最深之處具有內建的一些衝突或是
矛盾，使得到最後悲劇都不可能被避免，這也是可能的。不是
每一個問題都是有答案的。

不過在目前，這種臆測看起來似乎都是徒勞無功的。我們
連架構出恰當的問題都還太早，更不用說要去理解其答案了。
特別是我們腦部的結構可能需要先改善，這樣我們才有可能來
領悟宇宙的一些奧秘。

Eventually, most "philosophical" problems may turn out to be biological. Dr. Jonas Salk has written, "If we can study CNS [central nervous system] phenomena according to those biological principles that have been shown to be applicable to other systems, a basis for reconsidering behavior in biologically meaningful terms may emerge, which then by empirical means may expand further our understanding of the CNS of man and all that flows therefrom: behavior, creative activity, motivation, values, responsibility, and the intangible qualities of personality reflected in reactions, choices, aptitudes, and attitudes." (96)

When a humane, progressive, cooperative society has been achieved, the purpose of life will be learning and growth - the disclosure and then the attainment of ever more advanced intermediate goals, until either the final goal (if any) is revealed, or some catastrophe overtakes us.

During this grand unfolding, "happiness" in private and peripheral affairs will no doubt rest on a compromise between internal and external satisfactions. Few of us would want the contentment of a narcotic stupor, or an ESB jag, on a permanent basis. Likewise, ignorance may be bliss, but who would want the "happiness" of a cow - or even of a bull? Yet, judiciously used, chemistry and surgery can exert a beneficial stabilizing influence. On the external side, there will be an ever-widening range of activities, including some we have not yet thought about.

到最後，大部分的"哲學性"問題可能都會變成是生物性的。**Jonas Salk** 博士曾寫道，"假使我們能夠依據那些已經在其他系統中證實已經應用得宜的生物學原理，來研究 **CNS**(中樞神經系統)，一套以生物學尚有意義的辭彙來重新理解其行為的基準可能就會出現，因此藉著一些試驗性的方法，可能就可以進一步擴展我們對人類中樞神經系統的理解，以及由其中所衍生出的一切：行為，創造性活動，動機，價值觀，責任，以及在反應，抉擇，性向，和態度上所彰顯出來人格上的不可量度的品質。"（**96**）

當一個具有人性的，進步的，合作的社會已經達成之時，人生的目的將會是學習和成長——一些越來越進步的過度性目標會不斷地顯現，然後被達到，直到不是最終極的目標（假使存在的話）被彰顯出來，不然就是我們被某些大浩劫所毀滅，

在這個大解密的過程中，私人上和其週邊事務中的"快樂"，無疑地將會仰賴著在內部的以及外在的滿足間的一種調協。很少人會想要長期地靠著麻醉藥品的迷幻，或是一種腦部電子刺激（**ESB**）的激盪來獲得滿足感。同樣地，雖然傻人有傻福，但是想要那種乳牛似的"快樂"——甚或是一條公牛似的呢？然而，僅要小心謹慎地被使用，化學藥物和手術還是可以施加一種有益精神安定的效應。從外部來看，未來將會有越來越廣範疇的活動，其中還包括一些我們從未想像過的。

To some, this may appear remote and gray. Actually, the canvas of the future shows blinding color and riotous excitement. Remember, once again, you and I will not be the same, but enlarged and enriched, equipped fully to appreciate these words of Thomas Huxley: "If there is anything I thank the Gods for (I am not sure there is, for as the old woman said when reminded of the goodness of Providence - 'Ah but he takes it out of me in the corns') it is a wide diversity of tastes. . . No one who has lived in the world as long as you and I have can entertain the pious delusion that it is engineered upon principles of benevolence. . . But for all that, the Cosmos remains always beautiful and profoundly interesting in every corner - and if I had as many lives as a cat I would leave no corner unexplored." (45)

對某些人而言，這可能會有點遙遠而模糊。事實上，在未來的畫布上將散發出令人眼盲的光彩以及豐富盪漾的興致。請再度記得，你和我將不再復如往昔，而是會被擴展和豐富化，配備完整來體驗Thomas Huxley所說的話：“假如神有任何事情可以讓我感謝的話（我是不能確定有沒有，因為就好像當那個老女人被問及有關老天的恩典時所說的──‘感謝神，但是祂在玉米田中時就把它從我這兒拿走了’），其應該會是五味雜陳的。.........沒有一個像你我一樣活這麼久的人，會採納這一切都是依據天恩的原則所精心設計出來的虔誠的迷思。........但是儘管如此，宇宙在各個角落還是永遠保留著其美妙和深邃的興緻──而且假使我有九命怪貓一樣多條的生命，我一定會到處探索，不遺留任何角落。”（**45**）

CHAPTER X

Manners, Modes, and Morals of Tomorrow

明日的態度，模式和道德

CHAPTER X

Manners, Modes, and Morals of Tomorrow

Almost every commodity, however desirable or necessary, if it is not an immediate necessity, needs to be "sold," whether it is life insurance, food, medicine, or anything else. If respiration were not reflex, many people would have to be given a hard sell to draw a breath of air. Immortality itself will have to be "sold" to enough people to start the freezer programs.

Will tomorrow really be better? Is it worth the struggle? If we agree that it will become possible to mould personality, then logically this alone should assure an affirmative answer; we can all be made into jolly, as well as jolly good, fellows. Still, we would like some assurance that the external changes will be worthwhile also.

In the previous chapter we talked about the uses of immortality in the most general terms, and of course detailed prediction on a long-term basis is entirely out of the question. However, it will be amusing to spell out a few of the shorter-range possibilities, and it may help make the prospect seem more real and personal. There will be no attempt here to be systematic, let alone exhaustive.

第十章

明日的態度，模式和道德

幾乎所有的產品，不管是被想要的或是不需要的，假使其不是一種即時必要的，就須要被"行銷"，不管它是壽險，食物，醫藥，或是任何其它的東西。假使呼吸不是一種自然反射的話，有許多人要叫他吸一口空氣，就要對他進行強力的行銷。永生不死將必須要被行銷到足夠的人中，才有辦法開始各種冰凍的計劃。

明天真的將會更好嗎？這值得去努力奮鬥嗎？假使我們同人格意塑造在未來將會變成可能，那麼邏輯上，光是這個就足夠來獲得一個肯定的答覆了；我們都可以被變成是快樂的，並且是一些快樂的好人。但是儘管如此，對這些外在的改變將會是值得的，我們還是會想要有某些程度的保障。

在前一章中，我們以最普通的言詞討論到了有生不死的利用，但是針對其基於長遠上的細節預測，當然是完全不可能的。然而如果能說出幾個短期上的可能，將會是極有趣的事，而且這也會讓未來的前景看起來更為真實和切身。在這裡我們沒有嘗試要有系統性，更不用提到要能淋漓盡致了。

Before beginning this sketchy catalogue, it might be well to give just a hint of the more distant future - not by indicating anything that will be in it, for this is hopeless, but by specifying something that will not.

Beyond Beowulf

A favorite bromide of writers is that "human nature doesn't change." But the manifestations of human nature vary rather widely with cultural differences, as we know, e.g., from studies of identical twins reared separately; and soon there will be changes in its biological basis as well, with results beyond guessing.

It is true that we still read Beowulf, and the Iliad, and Hamlet, and many scholars blithely assume that these and similar works will remain in our culture forever. But in the last thirty or forty thousand years, the supposed tenure of modern man on earth, cultural changes have been relatively small, and biological changes virtually nil. In the next few centuries, the changes will be incomparably greater.

I am convinced that in a few hundred years the words of Shakespeare, for example, will interest us no more than the grunting of swine in a wallow. (Shakespearean scholars, along with censors, snuff grinders, and wig makers, will have to find new, perhaps unimaginable occupations.) Not only will his work be far too weak in intellect, and written in too vague and puny a language, but the problems which concerned him will be, in the main, no more than historical curiosities.

在還沒有開始這個項目的素描之前，先給大家有關較遠未來的一個提示也無傷大雅——此並非要指出在未來將會有什麼東西，因為這是不可能的，而是要點出未來將不會有的一些事物。

超越狼人

作家們的愛用的一個口頭禪就是"江山易改，本性難移"。但是就我們所知，人類本性的彰顯會因著文化上的差異而有極大的變化，譬如，從同卵雙胞胎分別養育的研究可看到；而且不久之後也會在其生物體上產生變化，而其一些結果則超乎猜想。

其實現在我們還是在閱讀狼人，依里雅得以及哈姆雷特，而且許多的學者都戲謔地認為，這些東西以及類似的作品將永遠會存在在我們的文化當中。但是在過去的三四千年中，假設中地球上現代人類的存在時間，文化上的改變都是相對上地短少，而且其生物上的變異也幾乎是沒有。然而在接下來的幾個世紀，這些變異將會是無法比擬地大。

我深信在幾百年之內，例如像是莎士比亞的言詞，只會像是豬隻嬉戲翻滾的叫聲，我們對其將不再會感到興趣。（研究莎士比亞的學者，以及片檢人，磨煙商，和假髮商，都將要去找一些或許是無法想像的新工作。）不僅是他的作品會變成有點心智薄弱，用來書寫的文字太過於曖昧和弄巧，而且他所關切的一些問題，大致上說來將不過是一些歷史中的怪事而已。

Neither greed, nor lust, nor ambition will in that society have any recognizable similarity to the qualities we know. With the virtually unlimited resources of that era, all ordinary wants will be readily satisfied, either by supplying them or by removing them in the mind of the individual. Furthermore, if civilization will have survived that long amid the titanic forces available, it would seem that satisfactory modes of living and mutual accommodation must have been worked out. Competitive drives, in the inter-personal sense, may or may not persist; but if they do, it will be in radically modified form.

It is impossible to say whether most of us will be resuscitated before or after man has worked really drastic changes in himself. My own guess is that most of us now living will be frozen by non-damaging methods, and the reversal of aging will be easier than a complete redesigning and rebuilding of the brain and body, and we can therefore expect to awaken while people are still more or less human. Let us then cast a few glances into the middle distance, and try to perceive some of the facets of life in this period.

Stability and the Golden Rule

As already suggested, the prospect of immortality should provide a strong damper on rash and impetuous action and anti-social behavior. National leaders will want to preserve their own skins, and will be forced to take a much longer view. A temporary advantage will become unimportant.

不管是貪婪，或是情慾，甚或是野心，在那個社會中的，將不會和我們所認知的品德有任何可體察到的相似性。因著那個時代所具有幾乎是無限制的資源，所有一般的需要都很容易就會被滿足，不管是藉著供應它們，或是藉著從個人的心理中將它們移除。況且，假使在所有巨大的軍事力量中，人類文明還可以存活那麼久的話，那看起來似乎一些相當不錯的生存模式，以及彼此之間的互相容忍之道已經有被解決了。在人與人之間的一些競爭驅策力，有可能會，也有可能不會繼續存在；但是假使它們繼續存在，其也將會以徹底改變過的模式存在。

我們大多數的人到底是會在人們真正徹頭徹尾地改變他自己之前或是之後被復甦，這是不可能預先知道的。我自己的猜測是我們大多數的人將會用非破壞性的方法來被冰凍起來，因此老化的逆轉將會比一種腦部和軀體的完全重新設計和崇建要來得容易許多，所以可以預期我們可以在人們或多或少還像是一般人的時候被喚醒過來。接著讓我們來看一看一些中程的景象，並且試著先來瞧一瞧在此期間的一些生活上的片面。

穩定和金科玉律

就如同先前所指出的，永生不死的境界中針對狂熱和衝動的行為，以及反社會態度，應該有一種強而有利的遏止機制。一些國家的領導人將會要保全自己的面子，而將會被迫去採取一個較長遠許多的觀點。一種暫時性的優點將會變成是不重要的。

Everyone's life will depend on the steady functioning of the freezers, and hence on the reliability of economic and administrative institutions. No one will be excessively greedy, in the knowledge that soon he will be stiff and cold and at the mercy of his successors, whose good will he dare not endanger.

In the era of the freezers, and still more markedly when immortality has actually been realized, there will be very salutary effects on interpersonal behavior. Our actions will be strongly influenced by the realization that not only ourselves, but the other fellow also, will be around a long time. The people we meet in business life and in casual encounters of every kind can no longer be counted upon to fade away and disappear; instead, our paths may cross repeatedly in a long future dimly seen. All business becomes "repeat" business; there are no more one-shots.

The Golden Rule then becomes not an ideal but a necessity, and there may well occur a Golden Age of morality and ethics, with every man counting every other his friend and neighbor.

(Some wiseacre is sure to ask what happens when a masochist tries to apply the Golden Rule. But it is not claimed either that the Rule is always crystal clear, or, even if it were, that the disposition to apply it would automatically liquidate dissension, but only that the Rule is on the whole a good one, and its general application would be a large step in the right direction.)

每一個人的生命將會仰賴著冰凍器穩定的運作，而因此就也仰賴著著經濟和管理機關的可靠度。在知悉自己不久將會變成僵硬而冰凍而聽任其子孫們的擺佈，不敢輕易去危害到他們好的關係，所以沒有人將會是貪得無饜的。

在冰凍的時代裡，更顯著地尤其是當永生不死已經實質地實現時，在人與人之間的態度上將會有極其可佩的效應。我們的行為將會被不僅我們自己，而且還有很多人都會活得很久的認知，而受到強烈的影響。我們在生意場合中所認識的人和在偶然的際遇中所碰到各色各樣的人，不再可以被認為是會從人生中退出和消失；反而，我們彼此之間的行徑，在遙不可見的漫長未來中會一再地交會。所有的生意會變成是"重複的"生意；而不再會是一次性的生意。

因此金科玉律將會變成是一種必需而不再是空言，而且未來將可能會產生出一個道德和倫理的黃金時代，其中每一個人都把其他人視為是自己的朋友和鄰居。

（一些聰明的傢伙一定會問道，當一個怪胎試著要遵行金科玉律時將會發生什麼狀況。不但規律並沒有說一定要是清清楚楚的，或甚至就算是的話，一但採行要遵循它時，其中不合宜之處就會自動地化解掉，而且就整體而言金科玉律都是好的，其普遍地實行將會是朝正確的方向所邁出的一大步。）

Possibilities of Stagnation and Decadence

Speculating about the ways of immortals, some writers have worried about decadence engendered by the excessive caution and timidity of those who are potentially immortal but vulnerable to accident. It has been conjectured that society would be emasculated, that new ventures would cease, that every citizen might eschew risks of all kinds - even refusing so much as to use vehicles for fear of an eventual accident.

This kind of development seems to me highly unlikely. In the second place, medical art will necessarily be so far advanced that few kinds of accident could result in permanent death. In the first place, creative drives and competitive pressures will persist in some form, and can be depended upon to keep the yeast fermenting nicely. As always, those who refuse risks and challenges will probably sooner or later be trampled into the ground - perhaps in a humane and genteel way, but firmly. As the lyricist says, "It's not the earth the meek inherit, but the dirt."

Even exceptionally dangerous jobs are not likely to go begging. A worthy cause, high pay, and glory will find at least a few takers for a long time to come.

More serious and sinister is the threat of decadence through strange and sophisticated new forms of seduction. We already have large numbers of "TV bums" with bent spines, bloated bellies, and stupefied minds from endless hours of slouching, snacking, and staring. Will an ordinary man be able to withstand the temptation to sit in a corner all day and tickle himself with ESB?

停滯和退化的可能性

臆測一些有關不死的情形，有某些作家就會擔心，由於那些可能會不死但是卻容易遭受意外事件的人的過度戒心和膽怯，所可能產生的退卻。他們臆測社會將會被癱瘓，新的創新嘗試將會停歇，而每一個人民可能會去避開各式各樣的風險——甚至於會因為害怕總有一天會發生車禍，而強烈地拒絕使用交通工具。

這樣子的發展就我看來是非常地不可能的。較次要地，醫學的技藝將一定會大大地進步，讓極少類的意外有可能導致永久性的死亡。最重要地，創造力的驅策和競爭上的壓力將會以某種型式持續存在，而可以靠著此來讓整鍋飯繼續好好地燉煮。跟過去一直一樣，那些拒絕風險和挑戰的人，遲早將有可能會被踐踏在地上——可能會是一種有人性和溫和的方式，但是卻是卻確實實地。就如同詩人曾說，"溫良的人所承繼的不是地土，而是泥巴。"

就算是特別危險的工作也不可能會沒有人要作。只要工作正當，有高酬勞，而且有光榮，久而久之至少就一定會找到幾個人要作。

最為嚴重而且諷刺的乃是經由一些怪異的和精巧的新樣式的迷惑，所產生出的退卻的威脅。我們已經有一大堆的"電視懶蛋"，因著一小時接著一小時的鬆垮垮，吃零食，和看電視，而變得彎腰駝背，大腹便便和心思愚鈍。一個一般的人是否能夠抵擋得住整天都坐在一個角落，用腦部電子刺激，**ESB**，來搔爽他自己的這種誘惑呢？

What will happen when the circuits of the brain are well understood, and hallucinations of the most convincing reality can be made to order, so that a man can rent a tape, put on his Dreamie helmet, and experience the part of the hero in a romantic adventure?

No neat and easy answer seems possible here. In China there actually are derelict souls who spend all their time in opium dens, if they can manage it. But this kind of activity would seem to be more or less self-limiting, since no one can retire from the world altogether without having someone else to look after him and his affairs.

An Eye for an Eye

It has been speculated that in a Golden Age criminals will be "cured" rather than punished. This notion seems to me faulty, or at least dubious, in three respects.

First, we cannot yet say for sure that every criminal is actually sick. He may be a healthy man who has decided (perhaps rightly!) that his interests and those of society do not coincide.

Second, even if antisocial behavior invariably resulted from a specific, curable disease, it would still be necessary to impose punishment for its deterrent effect. It is true that in medieval England petty crime flourished in spite of cruel punishments, and that crimes of impulse and passion are difficult to deter, and that many criminals are repeaters; but without deterrence the situation would be much worse, and the proposition remains generally valid.

當腦部的電路被完全摸透，而各種最具體真實的幻象都可以依需求被製造出來，這樣一個人就可以租一個帶子，戴上他的夢幻頭盔，去經歷在一個浪漫冒險中的英雄角色，這將會發生什麼事呢？

在這兒是不可能有簡單俐落的答案的。事實上在中國是有一些失魂落魄的人，假使他們付得起的話，就會整天呆在鴉片窖子裡。但是這類的活動看起來或多或少似乎是有自身限制的，因為沒有一個人可以在沒有其他某一個人來照料他和他的事情下，來完全地從這個世界退隱。

以牙還牙

曾經有人臆測說在一個黃金時代時，罪犯將會被"治療"，而不是被判罪處罰。這種概念就我看來，在三個方面是有瑕疵的，或至少是可質疑的。

首先，我們還不能確定說每一個罪犯實質上都是生病的。他可能是一個健康的人，只是他認定（可能是大義凜然地！）他個人的利益和社會的利益並沒有相符相容。

其次，甚至於假設說反社會的行為一成不變地都是從一種特定的，可治療的疾病所導致的，這還是需要來對其加諸懲罰，以達嚇阻之效應。的確在英國的中古世紀曾經在嚴刑峻法之下，盜竊奸犯還是肆虐喧囂，而這種衝動性和熱衷性的犯罪是難以嚇阻的，況且有許多罪犯都是屢犯的；但是如果沒有去嚇阻，情況就會更糟糕，所以此建議一般而言還是有用的。

Third, the general psychological atmosphere, and the feelings of the victims of wrongdoing, may for a long time demand traditional ideas of "justice," including its aspect of revenge.

It will become possible, in some sense and under some kind of interpretation, at last to make the punishment fit the crime. Culprits may be made to suffer all that their victims have suffered, and to make complete restitution. Would-be adventurers, exploiters, tyrants, and rogues will be curbed by fear of society's revenge: they will have much more to fear than death. There will be no more attitude of "might as well be hanged for a sheep as a lamb." If a tyrant causes a thousand people to be half starved for a year, he might be punished by being half starved himself for a thousand years - perhaps with a couple of centuries off for good behavior, or such other modification as would not weaken the deterrent principle.

Sex Morals and Family Life

It was indicated in Chapter VII that birth control will almost certainly sooner or later become de rigeur, at least for a certain period in history, and that this would happen even without freezers and immortality, simply because of the problem of natural increase. The availability and general practice of birth control by methods less clumsy than those now in general use - e.g., pills for either man or woman - and the much smaller average size of families and much smaller fraction of children in the population, will have many effects in many areas of life. But another development, seldom yet discussed, will have even larger effects.

第三，一般心理學上的氛圍，以及罪行受害者的感受，在很長一段時間內都還是會有傳統"正義"觀念的要求，其中包含了報仇的成分。

在某些意念上而且經過某類的詮釋後，至少在讓懲罰適合其罪行上，將會是可行的。我們可以使罪犯來蒙受他的受害者所蒙受所有的苦難，而來讓一旦得以完全重整。那麼可能的冒險者，掠奪者，獨裁者，赤色份子都將會害怕社會的報復而被治癒：他們將會有比死亡還要多許多的事情來害怕。未來將不再會有"不如把山羊當羔羊吊死算了"的心態了。假使有一個獨裁者讓一千個人挨餓一年，那麼他自己就可能被判要挨餓一千年——或許因為其表現良好可以減刑幾百年，或是有類似這樣的其它減刑方法，但是不能減弱嚇阻的基本原則。

性道德和家庭生活

在第七章中就曾經指出，生育控制至少在歷史中的某一時段中，遲早幾乎一定會變成是稀鬆平常之事，而就算沒有冰凍人和永生不死這回事，這還一定會發生，只是因為自然的增殖所可能帶來的問題。比目前普遍使用——例如不管是男性或是女性的避孕藥丸，較不麻煩的生育控制方法的出現和普遍地使用，以及小了許多的平均家庭成員數目，和人口中少了許多的兒童區塊，將會在生活中的許多部份有許多的影響。然而還有其它目前還很少被討論過的發展，甚至於將會有更巨大的影響。

Research proceeds apace on techniques of "ectogenesis," or the raising of "test tube" babies in artificial wombs instead of allowing normal gestation in the body of the mother. It is also foreseen that it will be possible to produce a child with only one parent, which could be either a man or a woman. (119) Such a parthenogenetic child, since it would have the same genetic make-up as its parent, would in a sense be a twin as well as offspring.

Ordinarily, one presumes, a child will have two parents. Few of us are so vain as to desire our children to be duplicates of ourselves. But ectogenesis will certainly become the rule: when it becomes available, what woman will prefer the ordeal of "carrying" and delivering a child?

At present, of course, many women will not admit the ordeal is disgusting, and may even insist it is "beautiful." But this is obviously just a psychological trick, making a virtue of necessity. One might just as well claim our methods of waste elimination are beautiful.

There will, of course, be a transition period and a rear-guard of opposition to this practice. Its opponents will be the same kind of people as those who howled against the use of anesthesia in childbirth when this was first introduced, claiming it was "unnatural," and that it was "intended" for women to suffer as punishment for their sins, and that mother love would be reduced if the pain of childbirth were removed.

科學研究在"人工繁殖"的一些技術上，也就是在人造子宮中培殖出"試管嬰兒"，而不讓其在母親的身體中進行正常的懷孕，可以說是突飛猛進。我們也可以預見到靠單一上一代就可以產出子代的可能，不管是生自於一個男人或適一個女人。（119）這種單親基因的小孩，因為其基因的組成和其上一代完全一樣，因此就某一方面來說，他同時是一個雙生，也是一個子孫。

　　一般我們都會假設一個小孩一定有父也有母。我們當中很少人會虛榮到希望說我們的小孩是我們自己的複製品。然而人工繁殖法將一定會變成通律：因為當其變成是可用時，有哪一個婦女還會喜歡"帶球跑"和生產小孩過程中的慘痛經歷呢？

　　當然在目前還有許多婦女不會將此痛苦認為是噁心的，而且甚至於堅持說這是一種"美感"。然而很明顯地這僅僅是一種心理上的詭譎，讓必需變成一種美德。如果這樣我們大可以也將我們的廢物排泄方式認定為是美麗的。

　　在未來對人工繁殖的實施當然會有一段的過渡時期，也會有最後防衛者的反對。其反對者和當麻醉技術剛被引進時，那些叫囂反對在生產小孩中使用此技術的人是同一類的，他們認為這是"不自然的"，讓婦女蒙受痛苦的"天意"乃是對來她們所犯罪孽的一種懲罰，而且假使生產的痛苦被一除了的話，母親的愛也將會減少。

Fathers love their children as much as mothers, without carrying or delivering them, and there will be no loss in this respect when ectogenesis becomes the rule. But there will be profound social changes.

Essentially, motherhood will be abolished. A child will usually not have a father and a mother, but instead will have two "fathers," one male and one female. The word "mother" may or may not persist, but its essence - gestation and delivery - will be gone. (Nursing is already a nearly abandoned custom in many communities.)

The differences between men and women will then be at a minimum. Most of the present differences are cultural, and the difference in physical size is of little importance; there is more difference in size between certain races of man than there is between men and women of the same race. Women will obtain genuine equality in almost all spheres.

In sex relations, women may become universally the aggressors. After all, they are not as definitely limited in capacity as are men; and when women need no longer fear pregnancy, the traditional roles of "taker" and "giver" may be reversed.

On the other hand, full equality may be restored by the discovery of ways to give men unlimited virility.

These developments need not be regarded as especially alarming. Sex is only a part of life, and not the most important part.

父親愛他們的小孩絕不會遜色於母親，雖然他沒有懷胎或是生產他們，而且當人工繁殖法會變成通律後，這方面也將不會有任何損減。然而未來將會有深遠的社會變革。

主要地，媽媽的角色將會被廢除掉。一個小孩通常將不會有一個父親和一個母親，而將會有兩個"父親"，一個是男性的和一個是女性的。"母親"這個字眼可能會也可能不會繼續存在，然而其實質——懷胎和生產——將會全然消失。（在許多社群中，哺乳幾乎已經是一種被捐棄的習俗了。）

於是，男人和女人之間的差異將會變成非常微小。當今的一些差異大多是文化上的，而體格大小上的差異是不太重要的；某些不同種族中男人之間的體格差異，要比同種族中男人和女人之間的差異來得大。未來幾乎在地球各地，女人將獲得真正的平等。

在兩性關係中，女人可能會全面性地變成主動者。畢竟，在性能力上她們並不像男人一樣那麼確定地有所限定；而且當女人不再需要害怕會懷孕時，傳統中的"接受者"和"給予者"的角色可能就會被對換了。

在另一方面，十足的兩性平等可能要靠著發現能夠給予男性沒有限制的元陽之氣，才可以被再度恢復。

這些進展不需要被認為是特別地緊急。性只是生活中的一部分，而且不是最重要的部分。

Its problems form only a part of the enormous package of problems we must wrap up.

The relaxed sexual habits that may develop do not seem likely to eliminate family life nor demolish the institution of marriage. It may become customary to experiment more or less promiscuously early in life, and marry at a later time than is usual now. But most people will still want children. Even in the absence of children, marriage serves an important purpose, as we know from the many successful childless marriages, and still more from the many successful marriages of divorced and widowed people whose children are grown and gone.

Most people, sooner or later in life, want and need the stability, comfort, and security of a relationship which neither friends nor blood relatives can supply.

The Question of Non-human Intelligent Entities

Modes and standards of conduct and intercourse may have to be developed with respect to intelligent creatures other than human. The three outstanding possibilities seem to concern the dolphins, robots, and extraterrestrial life forms.

It has been ascertained that dolphins have brains larger and more complex than those of men. Some investigators believe the dolphins, despite their lack of hands and artifacts, may be truly intelligent, and can perhaps be taught to communicate. (60) If this is true, we may eventually have to share the planet with them, and perhaps with some of their cousins, the whales.

其中的一些問題，僅僅是構成我們所必需解決的巨大問題包袱中的一部分而已。

有可能發展出來較釋放的性習慣，看起來是不可能會來將家庭生活消滅，也不會摧毀婚姻的制度。在年輕時或多或少去放浪形骸地體驗，而在比現在通常較晚的時間再結婚可能是會變成習俗。然而大多數的人將還是會要有小孩。就算是沒有小孩，婚姻還是有其重要的目的，就如同我們從許多沒有小孩的婚姻中，以及還有從許多離婚和鰥寡人士，而他們的孩子們都長大離開了的成功婚姻中可以了解到的。

大多數的人在其生命中，遲早都會想要和需要一個關係中的穩定，舒服，和保障，而這並非是朋友或是血親所能提供的。

非人類智慧個體的問題

未來行為和交往的模式和標準，除了必須要依照人之外，可能還要依照智慧生命體來研擬出來。似乎有三個懸宕的可能性乃是和海豚，機器人，和外星球生命的型態有關。

我們已經確知海豚具有比人類還要大而且還要複雜的大腦。有些科學家相信海豚雖然缺乏手和手工藝作品，但是卻可能是真正地具有智慧，而且可能可以被教會來溝通。（**60**）假使這是真實的，最後我們可能必須要和它們，以及可能要和它們的表親們，鯨魚，共同分享這個地球。

With respect to thinking machines, the problem is much thornier. To begin with, even though the philosophical notion of dualism has been singularly unproductive and the dualists in more or less steady retreat, nevertheless the mind-body problem remains unresolved. And even if we forget about dualism, it remains conceivable that a "machine" made of meat and gravy may have modes of existence not available to a machine of tubes and wires. That is, it is conceivable - although I think it farfetched - that regardless of their problem-solving, decision-making, and goal-seeking abilities, machines will never be worthy of the appellation "living."

But if we can find an appropriate test for the first thinking machine, Adam MacElectrosap, and discover he really does have awareness and essential life, then we shall be faced with a tough moral problem in deciding whether to keep him enslaved. (MAChine, ELECTROnic, SAPient, of course.)

We may also face a tough practical problem in deciding whether it is safe to keep him at all, enslaved or not. The possibility, celebrated in many a gruesome story, that our creations may some day turn on us and overwhelm us, is a real one. Professor Wiener, for example, believes that machines may not inevitably remain subject to man. (101)

After all, intelligent machines will necessarily have some degree of independence, initiative, and unpredictability - this is inherent in their intelligence, and this is why they are of value. Can we hope to control such entities, which will be in many respects, and perhaps in most respects, be superior to ourselves? The answer is not obvious.

至於有關會思想的機器，這個問題就要棘手多了。首先，雖然哲學上二元論的觀念一直都沒有什麼特別的成果，而二元論主義者或多或少也持續地在撤退，但是這個有關心靈－肉體的問題還是沒有獲得解決。而就算我們不談二元論，但是還是可以想像得到一台由肌肉和肉汁所造成的機器，就有可能有由管子和線路造成的機器所沒有的存在模式。這就是說，可以想像得到──雖然我認為是有點扯得太遠──不管它們所擁有的問題解決，決策下定，和目標追求能力，機器將永遠不配被指稱為"生命"。

　　然而假使我們可以為第一台會思想的機器，亞當麥克（Adam MacElectrosap），找到一種恰當的測試方法，而且發現到他的確擁有知覺和生命的實質，那麼我們就會要面對一個棘手的道德問題，必需要去決定是否應該繼續奴役他。（MacElectrosap 當然是指 MAChine, ELECTROnic, SAPient, 機器電子智慧人。）

　　不管我們有沒有奴役他，我們也可能要面對一個困難的實際問題，那就是要決定留住他是否安全。在許多恐怖小說中特別顯著的，那就是我們所創造出來的東西，有一天可能會背叛我們而傷害我們。這種可能性乃是一個真實的。例如 Wiener 教授就相信，不可避免地，機器可能不會永遠保持順服於人。（101）

　　畢竟，智慧型機器將必定要有某些程度的獨立性，主動性，和不可預測性──這是在它們的智慧中所賦予的，而這也是為什麼它們是有價值的。對這些在許多方面，而且甚或是在大多方面都比我們自己優秀的實體，我們能否有希望去控制它們呢？這個答案不是明顯易答的。

On the one hand, it is by no means unknown for the inferior intellect to dominate the superior. Greek scholars were held as slaves by Roman farmers. In certain environments, a tiger can kill a man. And it is perhaps conceivable that a mind could be massive and brilliant, yet mild and submissive. On the other hand, there must remain an element of doubt. One suspects that, sooner or later, the greater mind will have its way.

It is clear that no rules, no restraints, no restrictions can be relied upon; and of course it matters not at all that the machine has no direct physical powers or access to weapons. The machine, if it is to function, must be allowed to communicate, and if it can communicate it can probably persuade, and that is all that is necessary.

To realize just how bad the situation could be, we need only reflect that we may not even know where our interests lie! The machine will know what is best for us, and what is best for itself, and what courses of action are appropriate for these respective goals; but we may know none of these things, and be forced to rely on the machine!

The remedy, as hinted earlier, may lie in coupling a human brain to the machine, either permanently or occasionally. If the circuits can be integrated so that the machine is only an extension and enlargement of the man s mind, then the situation may be under control. This would also represent a new level of life for the man, an experience we can hardly imagine.

在一方面，要讓智能較低者來主導較高者，這絕非是前所未聞的。希臘的學者就曾經被羅馬的農夫用來當奴隸過。在某些環境狀況下，老虎是可以殺死一個人的。而且一個心靈可以既寬廣又聰慧，然而卻是既溫和又柔順，這或許是可以想像得到的。但在另外一方面，對這個還是要保留一點存疑。我們會猜想得到，遲早這個較大器的心靈將一定會走他自己該走的路途的。

很清楚地沒有任何規律，沒有任何限制，沒有任何禁令是可以信賴的；而這機器沒有實質的權力或是門路來使用武器，所以這當然是無所謂的。假使要讓這機器來進入運轉，那就必須要能夠容許有溝通能力，而假使他是可溝通的，那麼他就可以被說服，而這就足夠了。

要能夠知道狀況可能會變得多糟糕，我們僅要去反思到我們可能根本不知道我們的福祉是在哪裡就可以了！這機器將會知道什麼對我們最好，和什麼對他自己最好，以及要達到這些相關的目標，什麼是最恰當的行動路徑；但是我們可能對這些事情一無所知，而因此必須被迫仰賴這個機器！

其補救方法，就如前面所暗指過的，可能就是在於將一個人腦，永久性地或是偶而地和這機器耦合在一起。假使其間的電路可以被整合在一起，讓這機器僅僅是一個人類心靈的延展或是擴大，那麼整個狀況就可以被掌控。這也將會是代表人類生命的一個新層次，一個我們幾乎都無法想像的經驗。

Finally, turning to the question of possible extraterrestrial life, we find a riddle of awesome proportions. Where is everybody? Since the known universe contains at least 100,000,000 galaxies, with each galaxy numbering from 100,000,000 to 100,000,000,000 or more stars, and since most of these are at least several billion years old, some scientists think life must have developed on a great many worlds, and that in fact intelligent life must exist right now on myriad alien worlds.

Yet it is not true that "science agrees" there must be intelligent life on many worlds; the consensus of science appears to be uncertainty. (57) There does seem to be fairly good evidence that many stars have planets, that many of these may be suitable for life as we know it, and that under suitable circumstances life will probably arise. But there seems to be a possibility that most planets suitable for life have no land surface. More important, there seem to be no very securely based calculations to find the probability of intelligence developing from life, or civilization from intelligence. Therefore we must not be awed by the fact that the universe contains probably over 1,000,000,000,000,000,000,000,000 stars; the probability of civilization having developed in one of these systems could easily be much less than one in 1,000,000,000,000,000,000,000,000.

If civilizations are common, then civilizations in advance of ours should also be common - but in that case, why have there been no visitors? I know of no convincing explanation. The three most common suggestions are:

最後讓我們將問題轉到外星球生命的可能,我們發覺這是一個駭世驚人的謎題。人都跑到哪裡去了?既然已知的宇宙至少包含了一億個銀河系,而每個銀河系中又有為數從 100,000,000 個到 100,000,000,000 個或是更多的星球,而且因為這些星球大多數至少都有好幾十億年的壽命了,所以某些科學家就認為生命一定會在許多不同的世界中發展,而且事實上現在就應該有有智慧的生命存在在許多不同的外星世界。

但是如果說"科學家同意"一定有有智慧生命在許多不同的世界,這是不真實的;科學界的共識看起來是不確定的。(**57**) 的確似乎有相當好的證據顯示出許多星球都具有行星,而就我們所知其中有許多是適合生命存在的,並且在適合的狀況下,生命是有可能出現的。但是大多數適合生命的行星,看起來似乎有可能都沒有土地的表面。最重要地,看起來似乎沒有一些基礎非常穩固的計算方法,可以用來找出從生命發展到智慧,或是從智慧到文明的概率。因此我們也不應該因為宇宙中可能包含有超過 **1,000,000,000,000,000,000,000,000** 個星星就感覺到驚訝;文明在這些系統中的一個發展出來的概率極可能輕易地就小於 **1,000,000,000,000,000,000,000** 分之一。

假使文明是稀鬆平常的,那麼那些超越我們的文明也就應該是非常普遍——但是如過這樣,為什麼到現在都還沒有訪客呢?我還不知道有什麼具說服力的解釋。有三個最普遍的說法是:

(1) Time is too vast; all our neighbors are either far behind us or far ahead of us in development, and in either case cannot be expected to traffic with us; (2) Space is too vast, and because of the limiting velocity of light, and perhaps unknown dangers, interstellar travel is forever impractical; (3) "They" exist, and know about us, but just don't give a damn, or are watching but not interfering. All three suggestions seem implausible in light of our own psychology and prospects.

As we indicated in Chapter VII, the Golden Age will bring essentially unlimited wealth, with matter and energy freely available, and organization, in the form of thinking machines, also virtually unlimited. Then surely we will either scout the universe ourselves, or send out drone vehicles to investigate and report. If we find life, we will monitor it and take over its guidance and development, either out of charity or wariness. We will not allow fellow creatures to stumble on in misery, or develop into threats. The size of the universe means nothing: we have all the stars to tap for matter and energy, and our thinking machines can propagate themselves to any necessary number.

Certain dark suggestions have been made about the fate of man, and of those others who have failed to visit us. Perhaps civilizations that reach a high technological level always destroy themselves. Perhaps the fundamental problems of philosophy have no solution, and the final reward of progress is only the fullness of the realization that nothing matters; after the fruit ripens, the next stage may not be super-ripeness, but rot.

（1）時間是太廣杳了；我們所有的鄰居們在進展上，不是落後我們太遠就是領先我們太遠，而在任何一種狀況下，都不可能期望其能與我們同行；（2）空間是太浩瀚了，而且因為光線速度的限制，以及可能有一些未知的危險，星際間的旅行永遠是不可行的；（3）"他們"是存在的，而且也知道我們，但是只是不理不采，或是在觀察而沒有要干擾。在了解我們自己的心理和觀念下，所有這三個說法看起來都不太可採信。

就如同我們在第七章中所指出的，那個黃金的年代將實質地會帶來無窮的財富，其中物質和能量都垂手可得，而且以思想機器型態存在的組織也幾乎是能力無窮。因此我們一定會不是我們自己親自來探巡這個宇宙，或就會派遣一些遙控無人的航空器來調查和報告。假使我們發現了生命，我們一定會監控他，並且不管是出於救助或是戒備，會去承繼其導引和進展。我們將不會容許這些生命夥伴們去遭遇到慘狀，或是去發展成一些威脅。宇宙的廣大是無所謂的：我們有所有的星球來供我們探取物質和能量，而且我們的思想機器也會自行繁衍到任何所需要的數目。

有關人類和其他那些沒有辦法拜訪我們的的命運，有人提出某些陰暗面的說法。當文明到達一個高科技的水平，或許往往就會摧毀它們自己。或許哲學上的基本問題是沒有解答的，而其進步的最終報償就僅僅是去完全地體會到一切都無所謂；當果實成熟後，其下一個境界並不會是超級的成熟，而是腐爛。

But such pessimistic thoughts as these are premature, to say the least. For the present, let us simply acknowledge that the mystery remains a mystery, and also that we may, in fact, be ourselves the universe's elder race.

In any case, we gaze at the night sky and see the stars like dust, and reflect: either we are all alone in this vast universe, or else somewhere out there are other thinking beings, whom we may one day meet. Either way, it gives one pause.

Some Near-term Developments

Even the richest men of earlier times lacked many of the things available to the ordinary American and European today. These include: fast communication, fast travel, relatively reliable justice, accessible information, reliable emergency services such as fire and police departments, efficient plumbing, weather forecasts, insurance policies, loans on reasonable terms, dentistry, air conditioning, out of season foods, eyeglasses, anesthesia, and many other kinds of medical services and medicines. Certainly happiness is not directly proportional to wealth, comfort, safety, and peace of mind, but there is nevertheless a correlation.

Likewise we today are virtual paupers, compared to what we will be as resuscitees. Many extremely important goods, services, and modes of living will be available that do not exist today; some of these have already been indicated.

但是如同這類的悲觀思想，不管如何至少這些是不成熟的。以當下而言，就讓我們單純地認知到，這個謎團還是一個謎團，而且事實上我們自己也可能是宇宙中較老的族類。

在任何情形下，當我們凝視夜空，看見繁星多如塵土時，而來反思：在這浩瀚的宇宙中，我們要不是孤獨地存在，就是遠在某些地方，有其他會思想的生靈，有一天我們可能和其相會。不管何者，這都會讓一個人屏息凝思。

一些近程的發展

在古代就算最有錢的人，也沒有當今一般美國和歐洲人所能享有的許多東西。這些東西包括：快速的通訊，快速的旅行，相對可靠的司法，可獲得的資訊，可靠的緊急服務，像是消防隊和警察局，有效率的管線系統，氣象預報，各類保險，條件合理的貸款，牙醫，空調，非當季的食物，眼鏡，麻醉術，和其他許多種類的醫療服務和藥物。當然快樂不是直接和財富，書是，安全，和心靈的平靜成正比的，但是不管如何還是相關的。

同樣地，比起未來當我們是復甦者的情況，當今的我們可以說是名符其實的貧民。有許多非常重要的物品，服務，和生活模式，這些當今不存在的東西未來都將會有；其中有些都已經被點出了。

In addition to the qualitatively new things, there will soon be much available which requires essentially no technological advances or breakthroughs whatever - which requires, in fact, nothing except more work, more production, more automation, more wealth, of a kind that already exists, and ordinary progress.

Cities may be weather-controlled, if necessary by covering the streets with retractable roofs; the air and the streets will be kept clean and sanitary. A half-inch of snow will not tie up traffic. Hay fever and other allergy victims will have relief.

Safety and law enforcement in cities may be greatly improved in several ways. Public places may be monitored by recorded television, to speed assistance and to preserve evidence. (For example, all vehicular traffic may be continuously filmed, unless the legislatures decide the infringement of privacy is an overriding consideration.) Homes and even individuals may carry small emergency signal units, which could summon ambulances, firemen, police, tow-trucks, freezer technicians, etc. This might be combined with wrist radios, in the manner of Dick Tracy.

Honesty in private and public employment may be promoted by the use of periodic routine lie detector tests, covering prescribed areas, to ensure that trust has not been violated, as a condition of employment. Nothing helps morality so much as removing temptation. Of course, the legislatures may decide that this is akin to forcing a man to testify against himself, and disallow it, especially since it might be required of the legislators as well.

除了這些有特質的新東西外，對那些基本上不需要有科技上的進展或是突破的各種東西——這些東西事實上不需要任何條件，很快就可以變得非常易得，除了在已經存在的事物上，以正常的進展速度，更加的工作，更多的生產和更高度的自動化，來創造出更多的財富就行了。

都市可能都會有氣候控制，假使必要的話可以用可收藏的屋頂來覆蓋街道；空氣和街道都將可以保持乾淨和衛生。一場深達半英吋的降雪將不會阻斷交通。稻草熱和其它過敏的患者將得到紓解。

都市裡的安全和執法可能有很多方法來大大地改善。公共場所可能會用電視錄像來監控，以能加快就源和能保存證據。（譬如，所有汽機車交通可能會有連續不段的攝影，除非立法單位認為對隱私權的侵犯是構成一個否決的考量。）家庭，甚至於個人可能都會隨身攜帶小型緊急訊號器，這個可以用來呼叫救護車，消防隊，警察，拖吊車，冰凍技術人員，等等。這個可能可以和 Dick Tracy 電影中的一樣，將其和腕部無線電結合在一起。

在私人或是公家單位聘僱中的誠實度，可能可對涵蓋一些特別選定的區塊，用周期性例行的測謊試驗來提升，以便能確保信任沒有被違背，此可以當作聘僱中的一個條件。沒有比將誘惑除掉更能幫助工作士氣了。當然，立法人士有可能決定這樣有點像是逼迫一個人來對自己做不利的舉證，因此不會允許如此，尤其是因為立法人士有可能也會被要求來接受這種試驗。

Full-coverage liability insurance may be available and compulsory, so that everyone will be financially responsible, and collectible in case he commits any kind of wrong. Those whose records indicate they are poor risks will be insured by the state, but their activities may be restricted.

The Department of Health, Education and Welfare in the United States, and similar agencies elsewhere, may take increased responsibility for family life and training. At present, children are produced and raised usually by unskilled labor; little human beings are at the mercy of ignoramuses and brutes. The children will probably not, except in extreme cases, be taken from the parents, since it seems generally agreed that even a good orphan asylum is worse than even a rather bad family. But heavy pressure will be exerted to force parents to educate themselves and qualify themselves as parents, and the children will be protected through some kind of routine inspection.

Justice will be more uniform, more reliable, and cheaper. The absurd system of punishment typified by "thirty dollars or thirty days" will be discarded. Jail may be used only for people who are physically dangerous, or who may do irreparable damage, and not for those guilty of crimes strictly against property or of technical offenses such as violation of the anti-trust laws. Offenses in the latter category may be dealt with by fines linked to ability to pay, with credit given if necessary, and by supervised probation or restriction of activities. The rules of evidence will be drastically revised and modernized to allow a more logical evaluation of probabilities. The "reasonable doubt" rule may be replaced by a formula based on percentage probabilities.

概括承受的責任險可能會有，而且會是強迫的，這樣一來每一個人都將要負起財務上的責任，而且萬一他犯了任何種類的罪過，都將可以索取得到錢。對那些紀錄上顯示出是風險極大的人，將會由政府來對他們提供保險，但是他們的活動可能就會要受到限制。

美國的衛生，教育和福利部門，還有在其他地區的類似屬性單位，在家庭的壽命和訓練上可能要負上更多的責任。在目前，小孩往往大都是由經驗不足的勞動人士所生產和養育；這些小生命都任這些無知和粗魯的人所擺佈。除了在某些極端的特例外，這些小孩可能將不會被從其父母那裡帶走，因為大家通常都同意就算一個很好的孤兒庇護所，也會比甚至是一個頗壞的家庭還要來得糟糕。然而將會對其父母施加極大的壓力，迫使父母去教育他們自己，使他們自己合格來當父母，而小孩則會藉著某種例行的檢查來受到保護。

司法將會更為公平，更為可靠，而且更為便宜。那種被譏為"花三十塊錢，不然就關三十天"的荒謬的懲罰系統將會被棄置。監獄可能只會被用來關那些有人身危險的人，或是可能會造成不可收拾破壞的人，而不會關那些僅是犯了侵犯財產罪，或是例如不遵守反托拉斯法的技術上犯罪的人。犯了後面這一類罪的人，可能就會依其付款的能力以罰款的方式，假使需要的話還可以給以信用貸款，並且用受監控的假釋或是限制活動的方式來處理。對待證據的法則將會被大幅地修正和現代化，使其統計概率的評估較為邏輯化。那種"合理的懷疑"的法則可能會被一種架構在概率百分數的公式所替代。

Our republic could be transformed into a democracy, or perhaps a weighted democracy, through electronics. Every home might have a voting machine attachment built into its TV set, capable of identifying citizens by their fingerprints or retinal patterns or whatnot, and able to record and transmit votes. With the awkward machinery of voting thus streamlined, it might become practical to submit every important issue to referendum. Conceivably, the machine might first test the voter, and allow him to vote only if he proves he understands the issue reasonably well. It is also possible, as previously hinted, that the one-man-one-vote rule may be modified, giving a man instead a variable number of votes, depending on such things as his knowledge and the degree to which the issue affects him. (Admittedly, such notions would raise complex problems - but so did the replacement of buggies by autos, and so do most advances. The problems must be met and solved, and not dodged.)

Transcontinental supersonic subways, with fares low relative to average income, will allow everyone holidays and vacations in the mountains, forest reserves, or on either shore. In town, similar systems will fractionate commuting time.

The dull and unpleasant jobs will either be eliminated by automation, or compensated by shorter hours or higher pay. It is even possible that before very long all citizens will be allowed a basic income just for breathing, although jobs would be available for the qualified and would provide additional income. Perhaps the one inescapable form of work, and the main duty of all citizens, will be participation in political processes.

我們的共和國藉著電子設施可能會被轉化成一個民主,或是可能是一種加權的民主體系。每一個家庭中可能會有一台附加的投票機器內建在其電視機中,可以藉著它們的指紋或是瞳孔的模樣或是其他的特異性來辨認公民,並且可以用來記錄和傳輸選票。當那種齷齪投票機制因此而被流線化之後,把每一個重要的議題付諸票投就有可能變成可行。可想像得到地,這台機器首先可能要檢驗投票者,並且只有當他證明他對此議題有相當地理解,才准許他去投票。而且就如同先前所影射的,那種一人一票的法則也有可能會被修改,讓一個人因著例如他的知識和此議題會影響他的程度等情事,而可以有可變數目的票權。(不可否認地,這些觀念將會引發出一些複雜的問題──但是用汽車來替代馬車也會,大多數的進步都會。這些問題是要被面對和解決,而不是去逃避。)

有了車票費用相對於平均收入是低廉的橫跨大陸超音速地下鐵,將可以讓每一個在各個山嶺,森林保護區,或是各邊的海岸度其假期或是休假。在城市當中,一些類似的系統將可以讓上班通勤時間減少到一點點。

枯燥無味的和不舒服的工作將不是會被去除就是會被自動化,不然就會用較短的工時或是較高的報酬來被補償。不會太久,老百姓甚至於只要活著,就將有可能被付給一個基本的收入,然而還是會有工作給那些資格夠的人做,並且會付給額外的收入。或許對所有的公民而言,唯一不可避免的工作型式,而且也是主要的義務,將會是對一些政治活動的參予。

Those who find the mid-twentieth century a little lacking may well take heart. We have hardly begun to live.

那些感覺到在二十世紀中葉是活得有點缺憾的人可能大可以感到欣慰了。我們幾乎是還沒有真正開始活呢。

CHAPTER XI

The Freezer-Centered Society

以冰凍人為中心的社會

CHAPTER XI

The Freezer-Centered Society

Besides being definitely feasible, the freezer-centered society is highly desirable, and in any case nearly inevitable. This can be seen by illuminating more brightly, or from slightly different angles, a number of aspects introduced earlier.

Inevitability of a Freezer Program

It is easy to perceive that a large-scale freezer program must inexorably develop, sooner or later, whether or not my degree of optimism becomes general, and whether or not my personal efforts exert much influence.

We recall that suspended animation of humans (by freezing alive, without serious freezing damage, so that the subject can be thawed out and restored to active life at any time) is generally agreed to be in the cards. So far as I know, not a single expert doubts that this will come about, although there are wide differences of opinion as to when the technique will be mastered. Estimates vary' from about five years on up; my general impression is that a consensus might point to success within the lifetimes of a majority of people now living.

第十一章

以冰凍人為中心的社會

除了是一定可行的之外，以冰凍人為中心的社會是大家極度想要的，而且從各方面看來幾乎是不可避免的。這事是可以藉著針對先前所簡介過的一些觀點，將之更加地闡明，或是從稍微不同的角度，就可以看得出來的。

冰凍計劃是不可逃避的

不管我自己樂觀的程度有沒有變成是普遍的，而且不管我個人的努力是否有產生許多的影響，遲早大規模的冰凍計劃將會如火如荼地開展，這是很容易就可以察覺到的。

我們還記得人類的活體休眠（藉著活體時的冰凍，沒有嚴重的冰凍破壞，而讓受冰凍體可以在任何時間被解凍，並且恢復成活的生命）乃是一般都同意是計劃中的事。據我目前所知，沒有一個專家懷疑這件事是否會成真，雖然在有關這項技術將會被把握的時候上，存在著頗大意見上的差異。估計上的差異從大約五年開始起跳；我約略感覺到其中有一個共識，乃是將其可能成功時指向大多數目前活著的人的有生之年內。

As soon as suspended animation is practicable, persons with incurable diseases will surely be frozen alive to await the time that cures are discovered. It can scarcely be doubted that this development, at the very least and latest, would provide the entering wedge for the freezer program.

It is also a common assumption of both laymen and experts that medical science will find means of extending human longevity, at least in moderate degree. It is not likely to come in the form of a simple drug injection, although this remains conceivable and hints in this direction crop up from time to time. For example, a Royal Oak, Michigan, veterinarian, Dr. Henry Raskin, has been reported experimenting on dogs with a drug developed in Rumania, called GH-3; results in apparent revitalization of aged dogs are said to range from fair to spectacular. (17) More likely, the treatment will be complex and will only follow much longer study, but optimism is not lacking. Dr. Joseph W. Still, of George Washington University, has written: "Aging may prove to be no more fatal or inevitable than smallpox, polio, pneumonia, or tuberculosis." (111)

Now consider the outlook of an aged person in failing health, sometime late in this century, or maybe not so late. Suspended animation will be available; substantially increased longevity for those already old may not yet be at hand but research will be very promising; technology will be booming and wealth increasing by leaps and bounds. Obviously, there will be a great temptation to take the cold sleep for a few decades, or until a specified amount of progress has been made.

只要當活體休眠一變成可行時，那些罹患無法治癒的疾病的人將一定會尋求活體冰凍，來等待療法被發現的那一刻。這個進展，不管多晚和多少，將會為冰凍計劃的提供其開啟的楔子，這幾乎已經是不容置疑的。

醫療科學，至少在某個中級的程度上，將會找出延長人類壽命的一些方法，不管是業餘人士或是專家，此也是其共同認可的一個假設。雖然這還是可想而知的，而且偶而就會有這方面的暗示迸發出來，但是這不像是會以一種單純的藥物注射的型式來出現。例如說美國密西根州 Royal Oak 的獸醫，Henry Raskin 醫生，他一直在提出其用一種在 Rumania 所開發出來稱為 GH-3 的藥物，來在狗身上實驗的報告；其在老狗身上所產生可見的再度活化程度，據報是從不錯到相當可觀。(17) 這種療法更有可能地將會是複雜的，而且將一定要根據於更為長久的研究，但是絕非是缺乏樂觀的。美國喬治華盛頓大學的 Joseph W. Still 博士寫道: "老化可能會被證明是不會比天花，小兒麻痺，肺炎，或是肺勞更具有致命性或是更不可避免。" (111)

現在就讓我們來思考在本世紀末，或是說沒有那麼晚的一個健康垂危的老人所面臨的狀況。那時將會有活體休眠；雖然大幅度增長那些已經是老人的壽命可能尚未完全掌握，但是其研發將會是非常樂觀的；科技將會是突飛猛進的，而且財富將會跳躍式的增加。很明顯地，那時將會有很大的誘惑來接受冰凍睡眠幾十年，或是直到某一特定的科技進步程度已經達到時。

On awakening, this man and his wife can anticipate at least some added decades of active life in a more advanced world; in addition, compound interest will put him in a better financial situation. Why not sleep a seeming moment, and wake to a longer, brighter day? Who would not trade a few declining years in the present for a larger number of more active and rewarding years in the future?

Many, perhaps, would not -- but certainly many would. Some will make this choice, and others will follow, and finally it will become customary if not universal. Whether it comes soon or whether it comes late, whether the aim is "immortality" or something more modest, a large-scale freezer program is certainly going to mount, a majestic and irresistible tide.

Whoever would play the misguided and pathetic role of Canute, let him then he warned: he can only suffer dampened dignity.

No Generation of Martyrs

Since there is going to be a freezer program anyway, and since the frozen will share the immortality of their descendants, the rationale of opposition, if there ever was any, evaporates. Both immortality itself and the preliminary freezer program will bring their weighty problems, or exacerbations of old problems, but these can only be solved and not prevented.

在醒來的時候，這個人和其老婆至少將可以預期到可以在一個較為先進的世界中，生龍活虎地多活它幾十年；除此之外，因著複利的關係，他的財務狀況將會變得更有優勢。為什麼不來睡一看似短暫的片刻，而來在一個較長，較明亮的日子中醒過來呢？誰不會想去用目前一小段的風燭殘年，來換取未來更長更有活力而且更值得活的歲月呢？

或許會有許多人不會想要——但是一定會有需多人想要。某些人將會做這個抉擇，而其他的人將會跟進，而到最後這如果沒有變成是一種必然的話，也將會變成是一種習俗。不管其來臨是早，或是其來臨是晚，不管其目標是"永生不死"或是某些較為含蓄的目標，一個大規模的冰凍計劃將一定會躍起，而且是一股尊貴和不可抗拒的潮流。

任何想扮演那個迷失的和可憐的克努特 (Canute) 國王角色的人，就必須要被警告說：他是一定會蒙受到尊嚴的折喪的。

不需殉道者的世代

既然未來無論如何一定會有冰凍計劃，而且既然冰凍人也將和其子孫共享永生不死，假使有任何反對者的話，其反對的理由將會從人間蒸發。永生本身以及初期的冰凍計劃將會帶來他們所偏重的一些問題，或是使得一些老的問題更為嚴重化，但是所有這些都一定是要被解決的，而不是去避免的。

If by some stretch of the imagination a determined and concerted opposition to an early freezer program should cohere, its utmost effect could be to deny immortality to our own generation. A more monumental exercise in futility and sheer stupidity would be hard to conceive.

When an initially adverse reaction to the freezer idea is voiced, no matter what "reasons" may be given, it is usually based on nothing but pure funk. The idea unsettles people; it makes them nervous; it disturbs the established order; it raises questions and demands decisions. To many, especially those long beaten down by adversity, nothing is so precious as the "security" of a fixed routine and a known end; it is notorious that in the death camps of Nazi Germany many inmates refused any risk, preferring certain death to exertion.

Ostensible reasons for opposition often include various forms of asserted altruism. "We shouldn't burden later generations." "The future doesn't need us; I wouldn't want to live on unless I could do some good." "The money freezers would cost should be spent on cancer research or longevity research." "I'd rather a year were added to the life of a cancer victim than hundreds of years to my own." (The last two, of course, are non-sequiturs.)

Such self-styled altruists, who would martyr our generation, understand neither society nor themselves.

假使稍微伸展一下想像力，針對早期的冰凍計劃，萬一有一個堅決和集中的反對力量凝聚起來的話，其最大的效應將會導致對我們自己這一代永生的剝奪。那麼其中更為浩大爭取運動的慘敗以及道道地地的愚蠢，將會是無法去想像的。

當一開始一個對冰凍人觀念反對的反應被發聲出來時，不管其所仗的"理由"是什麼，其實它通常都不過是出於純粹的恐懼退縮。冰凍觀念會讓人們感到不安；它會讓他們精神緊張；它會擾亂到既有的社會秩序；它會製造出一些問題，而且必須要有所抉擇。對許多人而言，尤其是那些長期被逆境所打敗的，沒有什麼東西是會比一個固定例行的生活和一個已知結果中的"安定"更為彌足珍貴；在納粹德國的死亡集中營中，大家都知道有許多被囚者也都會拒絕去冒任何風險，甚至會為不想體力勞動而去選擇某類的死亡。

這些似是而非的反對理由往往都包括了各種宣稱是大公無私的樣式。"我們不應該加重後代的負擔。""未來並不需要我們；除非我能夠做一些好事，否則我不會想繼續活下去。""冰凍人所會耗費的金錢應該要花在癌症研究或是長壽研究上面。""我寧願讓癌症患者可以多活一年，也不願意讓我自己多活幾百年。"（最後兩個理由當然是無的放矢的。）

這些自我塑造出來的利他主義者，他們是我們這一代的志願殉道者，對社會或是他們自己一點都不了解。

We may be largely the intellectual heirs of the Greeks, but our moral heritage is Judeo-Christian, and in this tradition no babes are exposed on hillsides nor thrown to the wolves, no grandfathers are abandoned to die on the trail. We risk a division to rescue a battalion; we carry our wounded with us. We recognize duty downward as well as upward, from the state to the individual as well as conversely.

In fact, the worship of the State, or the Race, or Society, or Posterity, is merely a twisted and senseless sentimentality characteristic of totalitarian ideologies; it is nothing but fanaticism. In an important sense, there is no such thing as the state, no such thing as posterity: there are only individual people, and the living deserve as much consideration as the unborn. When someone who wouldn't give an extra hundred tax dollars to save a real, starving Indian claims he would sacrifice his life to make things easier for some hypothetical descendant, he is merely making an ass of himself.

In any case, of course, the direct remedy to the "burden" problem is easy: let us practice industry and thrift, so that the money for freezers is either extra money produced by extra work, or else savings diverted from fripperies. We can pay our own way, and need not be mendicants. Our estates and trust funds, through their investments and administrators, will contribute to future production and will share in control of the means of production. While we owe a moral debt to the future, the future will owe us not only a moral but a legal debt.

我們大致上可能可以說是希臘人聰明的後代，但是我們道德上的承繼卻是屬於猶太-基督徒的，而在這個傳統中，沒有嬰孩是會被丟棄在山腳下或是被拋棄在狼群中的，也沒有祖父母會被棄顧，而任其屍寒於道上的。我們都會冒著一師兵力的風險來拯救一個戰鬥蓮；我們都會背負我們的受創士兵。我們都任知道下層以及上層中的任務，從整個國家到每一個個人，逆向也是如此。

事實上，對國家，或是種族，或是社會，或是未來世代的崇敬，僅僅是在極權統治思維下的一種扭曲的，而且是荒唐的感性上的特質；其實這不過是過度的狂熱主義而已。從一個重要的觀點看來，其實本來就沒有所謂國家的東西，也沒有所謂未來世代的東西：而只是有單獨的個人以及所有活著的人，他們都需要和對還未生出來的人一樣多的照料。當某一個人不願意捐出多餘幾百塊錢的稅金，來救一個活生生正在挨餓的印度人，而卻宣稱說他願意犧牲自己的生命，來讓某些假想中的後代活得更容易一點，他簡直是在使自己變成一個王八蛋。

當然不管如何，直接解決所謂"負擔"問題的方法是容易的：讓我們過得勤勞而又節約，使得用在冰凍人的錢是來自於更加勤勞所賺的額外的錢，或是從無味的花費所轉省過來的儲蓄。我們可以付自己的花費，而不需要去化緣乞討。我們的家產和信託基金，藉著它們的投資和管理者，將可以對未來的生產有所貢獻，而且也將分享到生產工具的控制權。雖然我們是欠未來世代一個道德上的債，但是未來世代將欠我們的不只是一個道德上的，也是一個法律上的債務。

As to our "usefulness" in the future, it has already been pointed out that after resuscitation and rejuvenation we will be just as educable and adaptable as anyone else, young or old.

After maybe forty thousand years of struggling through the wilderness, the race has arrived at the banks of Jordan. Crossing will not be easy, nor will life in the Promised Land. But to pitch camp on the near shore for a generation would be a bootless waste.

It seems nearly certain that most of us will either see the point or will be initially in doubt. At first a few, and then mounting numbers will choose freezing, and before long only a few eccentrics will insist on their right to rot. Most people will not dare be left behind. There will be no generation of martyrs.

The Long View as Panacea

Well worth repetition, emphasis, and elaboration is the startling transformation in human relations which the freezer program will gradually work.

Not so long ago Sydney J. Harris, a syndicated columnist, remarked the effect on many people of the realization that we only live once. " 'I shall not pass this way again.' Then why does it matter what I do? Why not ruin the fields, deforest the woods, litter the roads, pollute the streams, trample the flowers, and treat people as a mere means to one's own ends?" (39)

至於我們在未來世代中的"有用性"，我們已經有指出，在復甦和回春後，我們將會和任何一個人，不管是老是少一樣地，可受教育和具有適應性。

人類在經過大約有四千年和蠻荒的搏鬥之後，終於來到了約旦河的河畔。跨河將不會是一件易事，而要能夠活在應許美地也將是不易。但是去紮營在近岸一個世代將會是一個無謂的浪費。

我們大多數的人看起來幾乎一定可以理解這一點，不然就會有一段初期的疑惑。首先會有幾個人，接著就會有越來越多人會選擇進行冰凍，而不久之後就僅剩下一些怪胎，才會去堅持他們所擁有可以去腐爛的權利。大多數的人將不敢去落於人後。未來將不需要有一代成為殉道犧牲者。

以遠見當萬靈丹

非常直得重複，強調，和深思的乃是在冰凍計劃所將逐漸作用下，在人際關係上所產生熾熱的轉變。

不太久之前，一個報系的專欄作家**Sydney J. Harris** 曾經對我們僅能夠活一次的認知，提出了其對許多人的影響的看法。"'我將不再經過這路途，'那麼為什麼要去介意我做什麼呢？為什麼不能將田園毀掉，將樹林砍光光，將路上丟滿垃圾，將所有河流污染，將所有花朵踐踏，並且以為追求目的而不擇手段來對待人們呢？" (**39**)

Although Harris was making a different point, it is obvious that a man who expects to be around for centuries or millennia will tend to behave differently from one who anticipates scant decades. In the long view, the fields, woods, roads, streams and flowers are my own; I cannot waste resources because I myself will need them later. I cannot cheat or injure a stranger, I cannot disregard his rights and feelings, because there are no more strangers, but only neighbors whom I will have to look in the face, again and again.

It has been fashionable for some time to say that "complex problems do not have simple solutions"; this is a favorite excuse of lack-wit politicians. Nevertheless, the simple use of soap and water cuts a very wide swath across the complex problem of disease prevention, and the simple routine of formal courtesy does wonders in ameliorating complex problems of human relations. Likewise, I believe the freezer program will prove virtually a panacea, particularly in international relations - not because in itself it solves all problems, but because it provides time for the solution of problems.

With an unlimited future to redress the balance, everyone can put up with temporary burdens and inequities patiently, if not cheerfully, and negotiate in good will. We all have a long, long way to travel together. When tempted to some rash action, one need only say to himself, "The end is not yet. The end is not yet. The end is not yet."

雖然 Harris 是在講另外一個不同的重點，但是很明顯地，當一個人預期他會活上幾百年甚或是幾千年，他一定會傾向於會和預期只能活少數幾十年的人有迴異的作風。以長遠的眼光看來，這些田園，樹林，道路，河流和花朵都是我自己的；因為我自己之後一直都需要它們，所以我不能浪費這些資源。我不能去欺騙或是傷害一個陌生人，我不能去忽視他的一些權利和感受，因為未來再也沒有陌生人了，而只會有我將必須一再地面碰面的一些鄰居。

有一句已經流行了許久的話說"一些複雜的問題是沒有簡單的答案的"；此乃是一些缺乏機智的政客所喜歡的一個推託之詞。然而，光是簡單地使用肥皂和水，在疾病預防上，就解決了這個複雜問題的一大半，而一些簡單例行的正式禮節，在改善人際關係間的複雜問題上，也都能產生奧妙的果效。同樣地，我相信這個冰凍計劃將會被證實幾乎像是一種萬靈丹，尤其是在國際關係上面——不是因為其本身可以解決所有的問題，而是因為它提供了時間來讓問題解決。

有了一段無限長的未來來修補其間的平衡，就算不是很愉快，但是如果以誠意來協調，每一個人都將可以耐心地忍受暫時的負擔和不公平。我們都將會有一段漫長的旅程會在一起。一個人如果忍不住要火爆三丈的話，他只要告訴他自己，"來日方長。來日方長。來日方長。"

All measures of desperation, including nuclear war, will tend to be ruled out. The reckless are usually those with little to lose - and there will be no more such, everyone will have a jewel beyond price - a glittering physical hereafter on the other side of the freezer. Heaven help Mao Zedong if he tries to persuade his people to turn their backs on this treasure, wrap themselves in tattered red flags, and lie down in moldy graves.

Time to Go Sane

Human life has always been based largely on fanatic lies and self-deception, a consequence of the endless struggle to solve the unsolvable, reconcile the irreconcilable, and scrutinize the inscrutable. Most of us have always preferred make-believe to frustration. But now at last it will be safe to go sane - at least partly.

The loyalties of the past have been mainly to ideas - usually stupid ideas, like the divine-right monarchies of post-medieval Europe, and often revolting ideas, like the blood-sacrifice rituals of the Aztecs. But the loyalties of the future will be to people - not disembodied abstractions, but individual human beings - and in this direction lies sanity.

Of course, in a sense it is only possible to be loyal to one's own thoughts, and in a sense other people are only thoughts. It is also true that doublethink and compromise with honesty will retain some utility. Still, the shift in viewpoint will be very real and very significant.

所有狗急跳牆下的手段，包括核子戰爭，都將可能被排除在外。那些亡命之徒通常都是那些無後顧之憂的人——但是未來將不再有這種人，每一個人都將擁有無價的珠寶——那就是在冰凍人之後的那個光彩亮麗的實存。假使毛澤東還試著要說服他的人民來拒絕這個寶藏，用破碎的紅旗包裹住他們自己，然後長眠在發黴的墳墓中，那只有老天幫得上他了。

該清醒的時候了

　　人類生命過去大都架構在一些狂熱謊言以及自我欺騙上，這是一個永無止境的掙扎，要來解決無從解決的，妥協無法妥協的，檢查無從檢查的因所造成的果。與其沮喪挫折，大多數的人通常都會因而選擇欺騙自己。但是現在我們終於將可以安全地選擇清醒了——至少部分地。

　　過去的忠誠主要都是針對一些思想上的——而且通常都是一些愚蠢的思想，像是歐洲後中世代君王的神聖權力，而且往往都是一些可惡的思想，像是阿茲提克的那些血腥犧牲儀式。然而未來的一些忠誠將會是針對人們的——不再是一些離體超現的抽象東西，而是一些活生生的個人——並且於此思想方向中蘊藏著清醒。

　　當然，就一方面來講，一個人僅有可能對他自己的思想忠誠，而另一方面，其他人都僅是一些思想而已。在誠實之下的雙重思想和妥協將還是會保有某些功能，這也是真實的。但是不管如何，未來在觀點上的轉移將會是非常地真實，而且是非常地顯著的。

We have usually thought of people as ephemeral, and ideas, especially "principles," as immortal. But now the people will persist while ideas come and go, and the results should be most salutary. Consider again the arch-villain Mao Zedong. Would he dare risk a fabulous life of thousands of years (including personal wealth eventually exceeding the total assets of the world today) for a moth-eaten bag of slogans and a shabby empire? Eternity, or some substantial portion of it, belongs not to Marxism-Leninism, nor to any other passing fancy in the mind of Mao, but to Mao himself and his relatives and friends -- including you and me. Once he understands this, he dare not risk war. If he cannot understand, those who do will remove him.

Fools, Madmen, and Heroes

Even after considerable thought, some people have to fight the feeling that to seek personal immortality is somehow ignoble, that the freezer-entered society is somehow distasteful and may rob us of our manliness. The reason is partly that bravery in the face of death has always been deemed a virtue, that abstract ideals are extolled above "selfish" ones, and that logic may seem to equate immortality with timidity. Even though the error of these notions has already been indicated, another remark or two will not be out of place.

我們通常都把人生想成是短暫過渡的，而把一些思想，尤其是那些"理論，"當作是永垂不朽的。但是現在人生將要變成長存的，而所有思想將會變成來來去去，這結果應該會是最有益健康幸福的。

　　請再想想那個大壞蛋毛澤東。他是否會敢去為了幾句老調長談的口號以及一個破碎的帝國，而來冒著喪失一個幾千年美好生命（其中包括其個人財富最終將超過當今世界所有資產的總和）的風險呢？永恆，或是其中的某大部分，並不屬於馬克思——列寧主義，也不屬於毛澤東心思中任何其他的突發奇想，而是屬於毛他自己以及他的親戚和朋友的——其中包括了你和我。一但他對這個有所理解，他就不會去冒戰爭的危險。而假使他不能理解，那麼那些能理解的人就會將他移除掉。

傻瓜，婦人和英雄

　　就算是經過深思熟慮之後，有某些人還是要去掙扎對抗那個認為追求個人的永生不死是有點不高貴，那個有冰凍人進入的社會是有點噁心而且可能會盜奪掉我們男性特質的感覺。其原因可能部分是來自於面對死亡的勇氣，過去一直都被捧吹為一種美德，這些抽象的意念被歌頌為超越那些"自私的"人，而且邏輯看起來似乎是把永生不死和懦弱劃成等號。雖然這些觀念中的謬誤已經被指出了，但是再提出一兩個說辭也不會不恰當。

Immortality is not an end in itself, nor do we reach for it in blind and breathless panic. It is an opportunity for growth and development otherwise impossible, and it is consistent with our highest current values.

The prospect of immortality will strongly color our lives, and in some ways dominate them, but it will by no means exclude other influences. We remain the products of our conditioning. I myself, for example, have been near death more than once, and would face it again without hesitation for any good reason, such as danger to my family or country.

We must ever bear in mind the gulf between the logical and the psychological. It has been noted that the long view will tend to rule out all measures of desperation; but some acts of madness or irresistible impulse will remain. On the other side, heroism will remain available not only because we are specifically trained for it, but because the subjective value of immortality, while large, cannot approach its face value. This is easily seen by remembering the behavior of Christians: in logic, nothing whatever is worth an eternity of hellfire, yet through the quirks of psychology countless millions are willing to be damned for the sake of paltry temptations.

Further, pondering of the problem of identity may convince some that extinction is nothing to worry about.

永生不死本身並非是一個目的，我們也不會盲目地和氣急敗壞地來追求它。其乃是一個成長和發展的機會，沒有了它一切就變成不可能，而且它是和我們目前最高的價值觀是一致的。

永生不死的前景將會大勢地讓我們的生命充滿了色彩，並且在某些方面會主導我們的生命，但是它將絕對不會排除其它的一些影響力。我們還會是我們自己調節下的產物。以我自己為例，我曾經瀕臨死亡超過一次以上，但是為了任何一個好的原因，例如對我家人或是國家的危害，我還是會毫不遲疑地去再度面對死亡。

我們對存在於邏輯上和心理上之間的鴻溝一定要永遠長記在心。前面已經指出，長遠的觀點將有可能來將所有絕望下的手段去除掉；但是一些一時瘋狂或是無法抗拒的衝動下的某些行徑將還是會存在。再另外一方面，英雄主義還是會存有，這不僅是因為我們已經特別地為此被馴練過，同時也是因為永生不死的主觀價值雖然是極為浩大，但是無法實現其表面實質的價值。從回想基督徒的行為，就很容易地可以看出其端倪：在邏輯上，沒有任何一種東西是值得去遭蒙永恆的地獄之火的，但是因著心理上的扭曲，無數千萬人竟會為了一些微小丁點的誘惑，而來甘心受到如此的咒組。

進一步，將有關身分上的問題經過深思熟慮後，就有可能來讓某些人相信，消滅其實是沒有什麼好擔憂的。

Finally, the steady workings of the process of natural selection will assure a continuing supply of heroes. A society without a sufficient percentage of risk-takers would scarcely be viable, let alone competitive.

These considerations also tie in with the misguided proposals that the freezer program be used as a eugenic sieve.

The Fallacy of Just-Freeze-the-Elite

One sometimes hears the naive asseveration, "Maybe we ought to save Churchill, but why should we save Joe Schmoe?"

The answer is easy, and comes in four parts:

1. Joe, after the future medicos work him over (although not necessarily immediately after resuscitation), will be just about as high-type and just about as useful as Sir Winston. He will no longer be the prisoner of his genetic inheritance.

2. If we are thinking in terms of rewards, perhaps Joe deserves first consideration, since Winnie has already licked a bushel of lollipops. Joe needs to be compensated for the sorry hand he was dealt the first time around.

3. The stratification of society is resented by the people in the lower strata. Even such trifling distinctions as those between master and slave, or between commissar and worker, are only grudgingly endured, if at all.

最後，自然選擇過程的穩定持續作用，將可以確保一些英雄持續的出現。一個社會如果失卻了一個足夠比例的敢冒險者，這個社會將幾乎無法存活，更不用去談競爭力了。

對一些誤導的提案，要將冰凍計劃用來當作優生學上的篩選工具，這些考量也會和其有所對應的。

僅冰凍精英者的迷思

我們有時候會聽到一種天真無知的假說，"或許我們應該去救邱吉爾，但是為什麼我們有必要去救陳進興呢？"

這個解答是簡單容易的，可以將之分為四部分：

1. 陳進興，在未來醫生把他整個處理醫療過後（雖然不一定是在其復甦之後馬上進行），將會幾乎是和邱吉爾溫斯頓爵士一樣地高尚和有用。他將不再會是他基因遺傳中的囚犯。

2. 假使我們是以報應來當思考模式，或許陳進興值得我們來優先考慮他，因為邱吉爾以經嚐過人生的甜頭了。陳進興反而應該因為他第一回合不幸所拿到的一手爛牌而受到補償。

3. 社會中的階級分層，一直都被在其底層的人士所憎恨。就算那些主人和奴隸之間，或是政治官員和工人之間的區別有多微小，如果有被吞忍下去的話，都還是心不甘情不願地。

The chance of the masses holding still for the vastly greater split between mortal and immortal is nil. The elite have a fairly simple choice: share immortality, or be torn limb from limb.

4. The benefits to all of society resulting from the long view depend on all of society sharing this view. The Golden Rule must know nothing of class or caste.

In short, the freezer program must embrace us all, with exceptions for minorities who voluntarily reject it. There will be a preliminary slipping and clashing of gears, but this must be kept to a minimum if the world's works are not to fall apart.

There is a saying: If the rich could hire people to die for them, the poor would make a good living. But our poor are not docile enough to be content with this kind of "living"; they will not build freezers for the rich, and then lie down themselves in slimy graves. Hence there must be no excessive time lag between the private, pioneer programs and public, mass programs.

Beginning of the Freezer Era 1964?

In a sense, the freezer era has already begun, since conscious, purposeful activity in this direction is under way.

社會大眾會因為在會死的人和不會死的人之間所存在，甚至於更為深廣的分隔，而來停擺的機會可以說是微乎其微。這些菁英份子有一種非常單純的選擇：那就是分享其永生不死，不然就要被五馬分屍。

4. 由於長遠的觀點所產生對整個社會的正面效益乃是有賴於整個社會對這個觀點的認同度。這個黃金定律一定要是不歧視任何的階級或是層次的。

簡而言之，整個冰凍計劃一定要能夠海涵我們每一個人，除了有少數一些人自願地要拒絕它的。在未來運轉的步調一定會有初期的打滑和衝撞，但是如果整個世界的運作不要因此而分崩離析的話，這種情形一定要被控制在最小的範圍。

有一段諺語說道：假使有錢人可以請人來替他們死的話，那麼窮人就可以過好日子了。但是我們的窮人並非如此順從到會滿意於這樣的"生活"；他們將不會為那些有錢人來建造冰凍庫，然後該該地自己躺在這個濕滑墳墓中的。因此，在一些私有的，先驅的計劃以及一些公共的，大型的計劃之間，一定不能有過度的時間延遲。

開啟冰凍人紀元−1964?

就一方面看來，冰凍人紀元其實已經開啟了，因為目前在這一方面的有意識，有目標的活動已經是在進行中。

There already exist, in late 1963, at least three organizations dedicated to furtherance of the freezer program, at least two of them legally incorporated. Many others can be expected to spring up shortly.

The freezer program is already a plank in the political platform of a congressional candidate, who now has the distinction of promising more than any other politician in history.

The grass-roots readiness, as indicated by my conversations and correspondence, is unmistakable - and oddly enough, it seems to have little or no relation to status or education; some poorly educated people are affirmative for the wrong reasons, and some scientists are against the program for emotional reasons. (There is wry humor in the predicament of any cryobiologists who may not favor the program; the poor devils will have to hope for their own failure!)

The first human may be frozen before the end of 1964, that is, within a few months of publication of this book. (Possibly a few wealthy people have been quietly frozen already!) Thereafter, events will gather speed, and our medical, financial, and political leaders may find themselves in the fix of Robespierre during the French Revolution. Robespierre, the story goes, was relaxing in a cafe with a friend when a howling mob went racing by. He jumped up and ran for the door. His friend called, "What's the matter? Where are they going?" Robespierre flung back: "I don't know where they're going, but I've got to get in front. I'm their leader!"

在 1963 年的後期，至少已經存在有三個組織，致力於冰凍計劃的推動，而它們其中至少有兩個是登記合法的公司。還有許多其它的組織預期在很短的時間內就會出現。

在政治性的政綱中，冰凍計劃已經成為是某一個國會議員候選人的政見，而他現在則會比過去歷史中其他的政治家有更多與眾不同的應許。

如同在我諸多的談話和書信中所指出的，民間草根中的預備程度是清楚可觀的──而且奇怪得很，這似乎和身分或是教育程度很少或是根本沒有關聯；某些教育程度不高的會因著錯誤的理由而來肯定，而某些科學家卻因著一些情緒上的理由來反對。（在任一個冰凍生物學家的矛盾中都存在著諷刺性的幽默，他可能不會贊同冰凍計劃；這些可憐的傢伙倒將必須要去希望他們自己是失敗的！）

在1964年年末之前，第一個人就有可能被冰凍，那就是在本書出版後幾個月之內。（也有可能有少數幾個有錢人已經悄悄地被冰凍了！）從此之後，事情將會增加速度，而我們的醫學界，財經界，和政治界的領袖們將會發現他們自己彷彿是被法國大革命中羅伯斯庇爾附身一樣。故事如此地進行，當有一群喧囂價天的群眾倉促經過時，羅伯斯庇爾正和一個朋友在一間咖啡屋中閒聊。於是他就一躍而起，朝著門口跑去。他的朋友叫說，"是怎們一回事呢？他們是要去哪兒呢？"羅伯斯庇爾倏然回身說："我不知道他們要去哪裡,但是我必須要去站在前面。我是他們的領袖啊！"

Hopefully, the freezer advocates will not have the less appetizing characteristics of a revolutionary rabble, but they will be just as determined. After all, the prize is Life - and not just more of the life we know, but a wider and deeper life of springtime growth, a grander and more glorious life unfolding in shapes, colors, and textures we can yet but dimly sense. Large numbers of Americans and Europeans will soon come not only to perceive but to feel the vastness and the grandeur of the prize, and to understand that all other prizes, all previous goals, are secondary. Their demands cannot be long ignored.

These demands will be of two general kinds, and will be aimed, among others, at physicians, biologists, morticians, insurance men, bankers, legislators, and lawyers.

First, make available routine and regularly updated procedures for freezing those now dying, making the most of current means.

Second, provide massive scientific and financial support for accelerated research in non-damaging freezing methods, as well as for a complete range of ancillary facilities.

In 1964, there will probably be little or nothing available in the form of institutional help or standardized procedures, and courageous individuals will have to take matters into their own hands. Then, for the first time in the history of the world, it will be *au revoir* but not Good-by.

很有幸運地,冰凍人的鼓吹者將不會有如一群革命群眾那種令人倒胃口的特質,但是他們將會有同樣的決心毅力。畢竟,其中的獎賞就是**生命**——而且不光是更多我們所知道的那種生命,而是一種更深邃更廣大,具有如春天成長力的生命,一種在各種型態,顏色,和紋理上所展現出的更偉大更光耀的生命,而我們現在只能朦朧地去感受到。多數的美國人和歐洲人很快不僅將可以來察覺到,而且還可以親身去體驗到這個獎賞中所蘊含的廣大和其雄偉,並且將會領悟到所有其它的獎賞,所有先前的目標都是次要的。他們的要求將不可能被長期地忽視了。

這些要求將會有兩個大類,而且除了會針對某些其它對象之外,將會特別針對醫生,生物學家,葬儀業者,保險人士,銀行家,立法諸公,和律師。

第一類,好好地利用現存的方法,使得要冰凍那些現在正要死亡的人,能夠有例行的以及常常更新的處理流程。

第二類,提供大量的科學上和財務上的支援,來加快在各種無破壞性冰凍方法,以及一個完整範圍的各類附屬設施的研發。

在 1964 年時,類似機構性的協助或是標準化的流程形式的東西,以及會自告奮勇的一些勇敢的人可能會很少或是根本沒有。從此之後,在這個世界歷史中的頭一遭,一切將會是*再見*,而不再是*好走*了。

REFERENCES
参考資料

1. Aldrich, C. K. "The Dying Patient's Grief." JAMA Journal of the American Medical Association, v. 184, no. 5, May 4, 1963, p. 329.

2. Asahina, E and Aoki, K. "Survival of Intact Insects Immersed in Liquid Oxygen Without Any Antifreeze Agent." Nature, Lond., v. 182, 1958, p. 327.

3. Asimov, I. The Chemicals of Life, New American Library New York, 1954.

4. Banfield, E. C. Quoted in The Detroit News, Feb. 18, 1962.

5. Becquerel, P. "La suspension de la vie au-dessous de 1/20 ° K absolu par demagnetisation adiabatique de l'alun de fer dans le vide le plus eleve. C. R. (Comptes Rendus) Acad. Sci., Paris, v. 231, 1950, p. 261.

6. Birkenhead, The Earl of. Famous Trials of History, Garden City Publishing Co., 1926.

7. Boerema, I. "An Operating Room With High Atmospheric Pressure." Surgery, v. 49, no. 3, March, 1961, p. 291.

8. Bogue, D. J. The Population Growth of the United States. The Free Press of Glencoe, Illinois, 1959.

9. Bradley, D. G. A Guide to the World's Religions, Prentice-Hall, 1963.

10. Brockman, S. K. and Jude, J. R. "The Tolerance of the Dog Brain to the Total Arrest of Circulation." Johns Hopkins Hospital Bulletin, v. 106, 1960, p. 47.

11. Bushor, W. E. "Medical Electronics, Part V: Prosthetics--Substitute Organs and Limbs." Electronics, July 21, 1961.

12. Cecil, R. L. and Loeb, R. F., editors. Textbook of Medicine, W. B. Saunders Co., 1951.

13. Constable, G. W. "Who Can Determine What the Natural Law Is?" Natural Law Forum, Notre Dame Law School, v. 7, 1962.

14. Corner, G. W. "Science and Sex Ethics." Adventures of the Mind (Second Series), Alfred A. Knopf, 1961.

15. Cscrepfalvi, M. Quoted in Health Bulletin, March 16, 1963.

16. Dahlberg, E. T. "Science and Religion at the Crossroads." Science and Religion. (J.C. Monsma, ed.), G. P. Putnam's Sons, 1962.

17. The Detroit Free Press, March 3, 1963.

18. The Detroit Free Press, March 19, 1963.

19. The Detroit Free Press, July 26, 1963.

20. The Detroit News, Jan. 24, 1963.

21. The Detroit News, May 27, 1963.

22. The Detroit News, June 16, 1963.

23. The Detroit News, June 17, 1963.

24. The Detroit News, June 19, 1963.

25. Edmunds, Folkman, Snodgrass, and Brown. "Prevention of Brain Damage During Profound Hypothermia and Circulatory Arrest." Annals of Surgery, v. 157, no. 4, April, 1963.

26. Egerton, N., Egerton, W. S. and Kay, J. H. "Neurologic Changes Following Profound Hypothermia." Annals of Surgery, v. 157, no. 3, March, 1963.

27. Elsdale, T. R. "Cell Surgery." Penguin Science Survey B 1963, edited by S. A. Barnett and Anne McLaren.

28. Evans, M. S. "The Compleat Growthman." National Review, v. 10, June 3, 1961, p. 352.

29. Feindel, W. "The Brain Considered as a Thinking Machine." Memory, Learning, and Language (ed. Wm. Feindel), University of Toronto Press, 1960.

30. Fernandez-Moran, H. "Rapid Freezing with Liquid Helium II." Annals of the New York Academy of Sciences, v. 85, 1960.

31. Fernandez-Moran, H. "Molecular Basis of Specificity in Membranes." Macrainolecular Specificity and Biological Memory (ed. Francis 0. Schmitt), The M.I.T. Press, 1962.

32. Fernandez-Moran, H. "New Approaches in the Study of Biological Ultrastructure by High-Resolution Electron Microscopy." The Interpretation of Ultrastructure (vol. 1 of Symposia of the International Society for Cell Biology), Academic Press, 1962.

33. Golay, M, J. E. "The Biomorphic Development of Electronics." Proceedings of the IRE (Institute of Radio Engineers), v. 50, no. 5, May, 1962.

34. Gorn, S. "On the Mechanical Simulation of Habit-Forming and Learning." Information and Control, v. 2, 1959, p. 226.

35. Gould, J. "Will My Baby Be Born Normal?" Public Affairs Pamphlet No. 272, Public Affairs Committee, 22 E. 35th St., N.Y., 1958.

36. Gresham, R. B., Perry, V. P. and Wheeler, T. E. "U. S. Navy Tissue Bank." JAMA, v. 183, no.1, Jan. 5, 1963, p. 13.

37. Haldane, J. B. S. "Life and Mind as Physical Realities." Penguin Science Survey B 1963, eds. S. A. Bamett & Anne McLaren.

38. Hardy, J. D. et al. "Re-implantation and Homotransplantation of the Lung." Annals of Surgery, v. 157, no. 5, May, 1963, p. 707.

39. Harris, S. J. The Detroit Free Press, May 22, 1963.

40. Health Bulletin, v. 1, no. 5, April 13, 1963. Rodale Press, Emmons, Pa.

41. Heinecken, M. J. God in the Space Age, John C. Winston Co., 1959.

42. Hotter, A. "Modification of Processes of Thought by Chemicals." Memory, Learning, and Language, ed. W. Feindel, University of Toronto Press, 1960.

43. Hotter, A. and Osmond, H. The Chemical Basis of Clinical Psychiatry, Charles C. Thomas, Springfield, Ill., 1960.

44. Holloway, M. R. An Introduction to Natural Theology, Appleton-Century-Crofts, 1959.

45. Huxley, T. H. "Letter to Sir John Simon," March 11, 1891, quoted by C. S. Blinderman in The Scientific Monthly, April, 1957.

46. Hyden, H. "A Molecular Basis of Neuron-Glia Interaction." Macromolecular Specificity and Biological Memory, ed. F. O. Schmitt, The M.I.T. Press, 1962.

47. The Insider's Newsletter, March 11, 1963.

48. The Insider's Newsletter, May 13, 1963.

49. The International Yearbook and Statesmen's Who's Who, 1962. Burke's Peerage Ltd., London.

50. Jacob, S. W. ct al. "Survival of Normal Human Tissues Frozen to - 272.2°C." Transplantation Bulletin, v. 5, p. 428.

51. John, E. R. "Studies of Memory." Macromolecular Specificity and Biological Memory, ed. F. 0. Schmitt, The M.I.T. Press, 1962.

52. Jordan, R. C. and Priester, G. B. Refrigeration and Air Conditioning, Prentice-Hall, 1956.

53. Kelly, J. L. Jr. and Selfridge, 0. G. "Sophistication in Computers: A Disagreement." Proceedings of the IRE, v. 50, no. 6, June, 1962, p. 1459.

54. Kemeny, J. G. "Man Viewed as a Machine." Scientific American, v. 192, no. 4, April, 1955, p. 58.

55. Kenyon, J. R., Ludbrook, J., Downs, A. R., Tait, I. B., Brooks, D. K. and Pryczkowski, J. "Experimental Deep Hypothermia." Lancet, ii, 1959, p. 41.

56. Kreyberg, L. "Local Freezing." Proceedings of the Royal Society of London B, v. 147, 1957, p. 546.

57. Krogdahl, W. S. The Astronomical Universe, Macmillan, 1962.

58. Kvittingen, T. D. and Naess, A. "Recovery From Drowning in Fresh Water." British Medical Journal, May 18, 1963.

59. Lillehei, R. C., Longerbeam, J. K. and Scott, W. R. "Whole Organ Grafts of the Stomach." JAMA, v. 183, no. 10, March 9, 1963, p. 861.

60. Lilly, J. C. Man and Dolphin, Pyramid Publications, 1961.

61. Lovelock, J. E. "Diathermy Apparatus for the Rapid Rewarming of Whole Animals from 0°C and Below." Proceedings of the Royal Society B, v. 147, 1957, p. 545.

62. Lovelock, J. E. "The Denaturization of Lipid-Protein Complexes as a Cause of Damage by Freezing." Proceedings of the Royal Society B, v. 147, 1957, P. 427.

63. Lund, G. "Is Faith Faltering Before the Scientific Advance?" Science and Religion, ed. J. C. Monsma, G. P. Putnam's Sons, 1962.

64. Lusted, L. B. "Bio-Medical Electronics-2012 A.D." Proceedings of the IRE, v. 50, no. 5, May, 1962.

65. MacDonald, D. K. C. Near Zero (An Introduction to Low Temperature Physics), Anchor Books (Doubleday & Co.), 1961.

66. Masse, B. L. "How Affluent Are We?" America, v. 101, Aug. 1, 1959.

67. Meier, R. L. Science and Economic Development, John Wiley & Sons, 1956.

68. Meryman, H. T. "Mechanics of Freezing in Living Cells and Tissues." Science, v. 124, 1956, p. 515.

69. Meryman, H. T. "Physical Limitations of the Rapid Freezing Method." Proceedings of the Royal Society B, v. 147, 1957.

70. Meryman, H. T. "The Mechanisms of Freezing in Biological Systems." Recent Research in Freezing and Drying, eds. A. S. Parkes and A. U. Smith, Blackwell Scientific Publications, Oxford, 1960.

71. Messenger, E. C. "The Origin of Man in the Book of Genesis" God, Man, and the Universe, ed. J. de Bivort de la Saudee, P. J. Kennedy & Sons, 1953.

72. Michigan Law and Practice Encyclopedia, v. 4, West Pub Co 1956.

73. Michigan Law Review, v. 23, 1924-25, p. 274 et seq.

74. Minsky, M. "Steps Toward Artificial Intelligence." Proceedings of the IRE, v. 49, 1961, p. 8.

75. Moore, E. F. "Artificial Living Plants." Scientific American, v. 195, no. 4, Oct., 1956.

76. Muller, H. J. "Man's Place in Living Nature." The Scientific Monthly, v. 84, no. 5, May, 1957.

77. Muller, H. J. "Genetic Considerations." The Great Issues of Conscience in Modem Medicine, Dartmouth Medical School Convocation, 1960.

78. Muller, H. J. "Life Forms to Be Expected Elsewhere Than on Earth." The American Biology Teacher, v. 23, no. 6, Oct., 1961.

79. Muller, H. J. "Mechanisms of Life-Span Shortening." Cellular Basis and Aetiology of Late Somatic Effects of Ionizing Radiation, Academic Press, 1962.

80. Neely, W. A., Turner, M. D., and Haining, J. L. "Asanguineous Total-Body Perfusion." JAMA, v. 184, no. 9, June i, 1963, p. 718.

81. The New York Times, May 19, 1963.

82. The New York Times Magazine, June 16, 1963.

83. Nigro, S. L., Reimann, A. F., Mock, L. F., Fry, W. A., Benfield, J. R. and Adams, W. E. "Dogs Surviving With a Reimplanted Lung." JAMA, v. 183, no. 10, March 9, 1963.

84. Olds, J. "Pleasure Centers in the Brain." Scientific American, v. 195, no. 4, Oct., 1956.

85. Page, R. M. "Man-Machine Coupling-2012 A.D." Proceedings of the IRE, v. 50, no. 5, May, 1962.

86. Pascoe, J. E. "The Survival of the Rat's Superior Cervical Ganglion After Cooling to -76°C." Proceedings of the Royal Society B, v. 147, 1957, p. 510.

87. Paul, J. "Culturing Animal Cells." Penguin Science Survey B 1963, eds. S. A. Bamett & Anne McLaren.

88. Pauling, L. "Chemical Achievement and Hope for the Future." Annual Report of the Smithsonian Institution, 1950, p. 225.

89. Penrose, L. S. "Self-Reproducing Machines." Scientific American, v. 200, June, 1959, p. 105.

90. Rey, L.-R. "Studies on the Action of Liquid Nitrogen on Cultures in Vitro of Fibroblasts." Proceedings of the Royal Society B, v. 147, 1957, p. 460.

91. Rey, L.-R. "Study of the Freezing and Drying of Tissues at Very Low Temperatures." Recent Research in Freezing and Drying, eds. A. S. Parkes and A. U. Smith, Blackwell, Oxford, 1960.

92. Rock, J. Quoted in The Detroit News, May 3, 1963.

93. Rosomoff, H. L. In Journal of Neurosurgery, v. 16, 1959, p. 177.

94. Rostand, J. "Glycerine et resistance du spenne aux basses temperatures." C. R. Acad. Sci., Paris, v. 222, 1946, p. 1524.

95. Rostand, J. "Can Man Be Modified?" Saturday Evening Post, May 2, 1959.

96. Salk, J. E. "Biological Basis of Disease and Behavior." Life and Disease: New Perspectives in Biology and Medicine, ed. D. Ingle, Basic Books, 1963.

97. The Saturday Review, "The Reversal of Death," Aug. 4, 1962.

98. Science Digest, "Frog Study Points to Possible Regrowth of Limbs," v. 48, Aug., 1960, p. 13.

99. Science and Math Weekly, American Education Publications, v. 3, issue 30, May 1, 1963.

100. Science et Vie, "La Vie centre Ie Temps," Yves Dompierre, May, 1963.

101. Science Newsletter, v. 79, no. 15, April, 1961, p. 234.

102. Science Newsletter, "Organ Exchanging Seen," v. 83, no. 17, April 27, 1963.

103. Scott, R. B. Cryogenic Engineering, D. Van Nostrand, 1959.

104. Sealy, W. C.; and Brown, Young, Smith, and Lesage. "Hypothermia and Extracorporeal Circulation for Open Heart Surgery." Anmils of Surgery, v. 150, 1959, p. 627.

105. Siekevitz, P. "Man of the Future." The Nation, v. 187, Sept. 13, 1958.

106. Simon, H. A. and Newell, A. "Heuristic Problem Solving: The Next Advance in Operations Research." Operations Research, v. 6, 1958.

107. Sinex, F. M. "Aging and Lability of Irreplaceable Molecules." The Biology of Aging, ed. B. L. Strehler, Waverly Press, Baltimore, 1960.

108. Sinex, F. M. "Biochemistry of Aging." Science, v. 134, no. 3488, Nov. 3, 1961.

109. Smethhurst, A. F. Modern Science and Christian Beliefs, James Nisbet & Co., Ltd., London, 1955.

110. Smith, A. U. Biological Effects of Freezing and Supercooling, Williams & Wilkins, Baltimore, 1961.

111. Still, J. W. "Why Can't We Live Forever?" Better Homes and Gardens, Aug., 1958.

112. Stone, Hrant, Donnelly, and Frobese. "The Effect of Lowered Body Temperature on the Cerebral Hemodynamics and Metabolism of Man." Surgery, Gynecology, and Obstetrics, v. 103, 1956, p. 313.

113. Strehler, B. L. Time, Cells, and Aging, Academic Press, 1962.

114. Taube, M. Computers and Common Sense, Columbia University Press, 1961.

115. Tenney, M. C. "Revelation." The Biblical Expositor, ed. C. F. H. Henry, vol. Ill, The New Testament, A. J. Holman Co., Phila., 1960.

116. Teuber, H.-L. "Perspectives in the Problems of Biological Memory - A Psychologist's View." Macromolecular Specificity and Biological Memory, ed. F. 0. Schmitt, M.I.T. Press, 1962.

117. This Week Magazine, "How a 'Frozen' Astronaut May Reach the Stars," Jan. 14, 1962.

118. Thomas, G. E. "Science and Religion -- a Partnership." Science and Religion, ed. J. C. Monsma, G. P. Putnam's Sons, 1962.

119. Time, "Biology of Individuality," v. 71, June 2, 1958, p. 47.

120. Time, "The High Cost of Dying," v. 71, June 2, 1961, p. 46.

121. Trevarthen, C. B. "Double Vision Learning in Split-Brain Monkeys." National Academy of Sciences, Autumn meeting, 1961.

122. Ullman, J. E. "Economics of Nuclear Power." Science, v. 127, no. 3301, April 4, 1958, p. 739.

123. Vallee, B. L. and Wacker, W. E. C. "Medical Biology - A Perspective." JAMA, v. 184, no. 6, May 11, 1963, p. 485.

124. Vital Statistics of the U.S., U. S. Department of Commerce, 1959, vol. II.

125. Walter, W. G. "An Imitation of Life." Automatic Control, Simon & Schuster, 1955.

126. Waterman, A. T. "Science in the Sixties." The Advancement of Science, v. 18, no. 72, July, 1961.

127. West, J. S. Congenital Malformations and Birth Injuries, Association for the Aid of Crippled Children, New York, 1954.

128. Wiesner, J. B. "Electronics and Evolution." Proceedings of the IRE, v. 50, no. 5, May, 1962.

129. Wolfe, K. B. "Effect of Hypothermia on Cerebral Damage Resulting from Cardiac Arrest." American Journal of Cardiology, v. 6, 1960, p. 809.

130. Wright, D. A. "Thermoelectric Cooling." Progress in Cryogenics, ed. K. Mendelssohn, v. 1, Heywood & Co. Ltd., London, 1959.

Ria University Press
www.ria.edu/rup

www.ingramcontent.com/pod-product-compliance
Lightning Source LLC
Chambersburg PA
CBHW021841020426
42334CB00013B/142